Cancer Exercise Specialist Studyguide/Handbook 8th Edition

Andrea Leonard, B.A., C.S.C.S., C.P.T, CES
along with
Dr. Glenn B. Gero – Conquering Cancer with Nutrition
Dr. Joseph Camp - Pilates

Table of Contents

General Information

This handbook is designed to help qualified individuals prepare for the CES workshop or home study. The program is designed for medical and fitness professionals who have a strong working knowledge of anatomy and physiology, and carry a national certification and/or degree in an exercise or health related field of study. This qualification process is designed to evaluate competence in the participants' knowledge, skills, and abilities found in the CES handbook.

Workshop Information

CES workshops are designed to develop and enhance the knowledge base and skills of the participants. The curriculum has been developed so that instructive material and its' practical application are scheduled together. This allows the participant to better assimilate both theory and practice.

In addition to the recommended prerequisites, participants should have adequate knowledge and background in the health or fitness profession. The workshops are not intended to provide the full experience and knowledge necessary for the CES.

Home Study Information

The home study program is available to those who are unable to attend a scheduled workshop. While the materials are very comprehensive, we strongly urge you to attend a CES workshop in the future to provide you with the critical hands-on information. You may attend any scheduled workshop within two years of registering for the home study program.

You do not have to wait to attend a workshop in order to take your examination. You may take it whenever you feel that you are ready. Upon passing of both the multiple choice and essay portions of the examination, you will receive your CES Advanced Qualification Certificate, CEU's, and be added to our online listing of International Cancer Exercise Specialists.

Qualification Information

CES qualification encompasses cognitive and practical competencies and is evaluated in both multiple choice and essay format during the examination. The candidate must successfully complete both components to receive the advanced qualification. Additionally, current cardiopulmonary resuscitation (CPR) certification is required to maintain the qualification.

Multiple choice examination – this portion of the examination includes 100 multiple choice questions. Questions are drawn from the CES handbook and lecture content.

Essay examination – this portion of the examination will attempt to evaluate the candidates' skills related to conducting sessions and evaluations. Although many acceptable protocols exist in the field of fitness and rehabilitation, the protocols used in this handbook were chosen as a representative sample of available, valid information and studies. **You will have exactly 6 weeks from the date of the workshop to complete your examination and have it mailed back to us, or there will be a $100 late fee.**

Results and re-tests – a candidate may be re-tested on the multiple choice or essay portion, or both if necessary. The results of the examination will be mailed to the candidate in approximately three to four weeks.

Candidates who fail either portion of the examination will need to pay a $100.00 re-test fee. Candidates will have six months from the original test date to successfully complete their qualification. Re-tests can be arranged by calling (503)502-6776 or e-mail at empowersurvivor@earthlink.net. Tests will be mailed to candidates' home address within approximately two weeks of receipt of payment. Candidate will have one month to complete the re-test and return it to The Cancer Exercise Training Institute.

Qualification Renewal

In an attempt to ensure ongoing competency and to maintain a high standard for qualified professionals, CES qualifications must be renewed **every two years.** The qualified professional will be given a two-month grace period after their qualification expires, in which to take the examination. CEU's will vary from certification to certification.

Cancellation policy

The following policy applies to workshop cancellations:

1. **The Cancer Exercise Training Institute has the right to cancel a workshop with two-weeks' notice due to low attendance. We are not responsible for reimbursing travel expenses that you may have incurred.**

2. If a workshop is cancelled due to low attendance, an attempt will be made to reschedule it for another date. If you would prefer to get a full refund, you must send your **unused** handbook back to CETI within one month of the original workshop date. Once it is received, a refund will be issued. You may also participate in the home-study program which will allow you to receive your Advanced Qualification and CEU's immediately and then attend any upcoming workshop free of charge within a two year period.

3. If you are unable to attend a workshop for any reason, you may attend any workshop scheduled within a two-year period, free of charge.

4. If you order the home study materials and are dissatisfied for any reason, you may send your unused handbook and DVD set back within one month of the original purchase date for a full refund. After one month, NO refunds will be honored.

Cancer Exercise Specialist Advanced Qualification

The Cancer Exercise Specialist (CES) is a professional qualified to assess, design, and implement individual and group exercise programs for individuals diagnosed with cancer. The CES is skilled in evaluating health behaviors and risk factors, conducting comprehensive fitness assessments, writing appropriate exercise recommendations, and motivating individuals to modify negative health habits and maintain positive lifestyle behaviors for health promotion. The CES will have a complete understanding of the entire cancer process from diagnosis to treatment, recovery, prevention of lymphedema, and contraindications. The qualification process includes demonstrating competency through a multiple choice and essay examination. It is granted to candidates who score eighty percent or higher on the examination.

Recommended pre-qualification competencies and work experience:

1. Work related experience within the health and fitness or medical field.
2. Educational training comparable to an undergraduate degree in health and fitness or a closely related field.

3. Adequate knowledge of and skill in risk factor identification, fitness appraisal, exercise recommendations, and basic nutrition.

4. Demonstrated ability to incorporate suitable and innovative activities that will improve an individuals' functional capacity in conjunction with their cancer treatment and surgical procedures.

5. Demonstrated ability to effectively counsel individuals regarding lifestyle modification.

6. Demonstrated competence in the knowledge and skills required for the CES.

7. Current CPR.

Course Description

Health and fitness professionals will expand their knowledge base by learning all stages of cancer treatment, side-effects, and reconstructive procedures and how they apply to and/or contradict exercise programming.

Objectives:

1. Students will be able to identify prevention, identification, and management strategies for lymphedema.

2. Students will be able to identify all contraindications to exercise during cancer treatment and/or surgery.

3. Students will be able to demonstrate proper techniques for completing an in-depth physical assessment on their clients.

4. Students will understand the physical implications after surgery and/or treatment of twenty-five types of cancer.

Sample Questions and Answers

The following sample questions will help the student assess his or her knowledge base in preparation for the multiple choice portion of the CES examination. The questions reflect the type of question asked and the depth of knowledge expected. The answers follow the last question. The student is encouraged to review, in detail, those topics for which his/her answers were incorrect.

1. The function(s) of the lymphatic system is/are:
   - a) aiding the immune system in protecting the body from disease
   - b) returning fluids to the blood in the circulatory system
   - c) transporting fat from the digestive tract to the blood
   - d) filtering bacteria, viruses, tissue debris, and other foreign substances from body fluids
   - e) all of the above

2. After an Abdominal TRAM procedure, particular attention must be paid to strengthening which of the following muscle groups in order to stabilize the torso?
   - a) obliques and intercostals
   - b) latissimus dorsi
   - c) erector spinae
   - d) both a and c
   - e) all of the above

3. A potential client forwards their medical release form to their doctor. The doctor will not give their approval for the client to participate at this time. However, the client still wants to participate. You should:
   - a) give them the proper paperwork and get them enrolled before they change their minds
   - b) suggest they be reevaluated by their doctor and check back with you at a later date
   - c) allow them to enroll, but let them know that they will only be allowed to do stretching and range of motion exercises until you have their doctor's permission
   - d) give them a some stretches to work on and suggest they do them for a few weeks before they go back to their doctor for reevaluation
   - e) both a and c

4. We currently know that lymphedema may be caused by:
   - a) radiation treatment to the lymph nodes
   - b) lifting weights in excess of 12 lbs
   - c) antidepressant use

d) drinking more than 8 glasses of water daily

e) none of the above

5. Patients undergoing chemo./radiation will reap the same physiological training results as those not undergoing treatment.

   a) true

   b) false

6. Which of the following is **not** a sign of cancer fatigue?

   a) lack of interest in normal day-to-day activity

   b) increased appetite

   c) increased time spent lying in bed or sleeping

   d) not having enough energy to do normal activities

   e) all of the above

7. The fingertip wall walk will help to do which of the following:

   a) increase ROM in shoulder abduction and flexion

   b) increase ability to externally rotate

   c) improve shoulder extension

   d) prevent lymphedema

   e) none of the above

8. With radiation therapy, pain is often associated with which of the following?:

   a) breakdown of mucous membranes

   b) scarring of the nerves (fibrosis)

   c) swelling

   d) skin sores

   e) all of the above

9. You are scheduled to begin training a woman with a slight case of lymphedema. Prior to your first workout you should do which of the following?

   a) contact their doctor for permission

   b) contact their physical therapist or lymphedema specialist

   c) instruct them to wear their sleeves or wraps as prescribed by their doctor or therapist

   d) both a and b

   e) all of the above

10. Patients should be advised **not** to have a flap reconstruction in which of the following situations?

a) smoking history of 15+ years

b) patient has diabetes

c) family history of breast cancer

d) when there is bi-lateral breast cancer

e) both a and b

11. A blood transfusion may be required if a cancer patients' red blood cell count drops too low.

a) true

b) false

12. To help to prevent the onset of lymphedema, a patient should practice which of the following precautions?

a) use moderation when working the affected body part

b) always get blood drawn or shots given in the unaffected arm or leg

c) elevate the affected arm or leg whenever possible

d) both b and c

e) all of the above

13. All of the following are road blocks to pain control **except**:

a) nausea

b) sedation

c) hallucinations

d) premature ventricular contractions

e) allergic reactions

14. When performing a postural assessment it is important to have your client stand still and maintain ***perfect posture*** while you conduct the assessment.

a) true

b) false

15. Prior to beginning or resuming a ***strength training*** routine after a mastectomy, your client should have 90% or better range of motion in her affected arm.

a) true

b) false

16. What precautions must be taken with a client who has undergone an extensive cervical lymph-node dissection for medullary thyroid cancer?

  a) don't let client overheat
  b) make sure that they stay well hydrated
  c) use moderation for all upper back and neck strength training
  d) emphasize stretching of the neck area
  e) nnnall of the above

17. Somatic pain can be described by which of the following:

  a) sharp pain
  b) aching
  c) throbbing pain
  d) b and c only
  e) all of the above

18. At which stage is lymphedema still potentially reversible?

  a) stage I
  b) stage II
  c) stage III
  d) stage IV
  e) none of the above

19. The five sub-classes of cancer, that are not carcinomas, include all of the following **except:**

  a) sarcomas
  b) melanomas
  c) leukemias
  d) lymphomas
  e) blastomas

20. The term "hidden cells" is used when referring to which of the following?

  a) cancer cells not detected at the time of surgery
  b) cancer cells that are found in the ducts or lobules of the breast
  c) cancer cells that have found their way into the lymph nodes
  d) cells that move through the blood stream, making them almost impossible to find
  e) none of the above

Answers to sample test questions

1. **E** / 2. **D** / 3. **B** / 4. **A** / 5. **B** / 6. **B** / 7. **A** / 8. **E** / 9. **E** / 10. **E** / 11. **A** / 12. **E** / 13. **D** / 14. **B** / 15. **A** / 16. **E**

17. **E** / 18. **A** / 19. **E** / 20. **A**

Essay Examination

The essay examination will attempt to evaluate candidates' skills related to conducting sessions and evaluations. Although many acceptable protocols exists in the field of fitness and rehabilitation, the protocols used in this handbook were chosen as a representative sample of available, valid information and studies. The candidate will be expected to describe, in detail, protocol for the following:

- **Assessment of shoulder range of motion**
- **Postural analysis**
- **Circumference assessment**
- **Exercise programming for various case studies**

In the case studies, candidates will be expected to include all potential contraindications, prevention measures, and exercise and lifestyle recommendations.

Examination scoring

The examination process should take approximately three hours. To pass the multiple choice portion of the exam, a score of eighty percent or better will be required. The essay portion will be graded pass/fail. Knowledge, content, and accuracy will be evalutaed in scoring questuons on the essay exam. Although some items will be weighted heavier than others, no item will be weighted so high as to allow a candidate to fail based soley on failure of that item. The candidate must fail a number of items in order to fail the examination. A failing score on either section of the examination will require re-testing in that area.

Lesson Plan for Cancer Exercise Specialist Workshop

Day One

9:00-9:15 **The disease/cancer classifications**

9:15-11:00 **Surgical procedures/treatments**

1. **Prostate**
2. **Testicle**
3. **Lung**
4. **Brain**
5. **Colon or rectum**
6. **Bladder**
7. **Kidney**
8. **Pancreas**
9. **Liver**
10. **Stomach**
11. **Small intestine**
12. **Ovaries**
13. **Uterus**
14. **Cervix**
15. **Lip and oral cavity**
16. **Thyroid**
17. **Esophagus**
18. **Larynx**
19. **Throat**
20. **Leukemia**
21. **Lymphoma**
22. **Multiple Myeloma**
23. **Bone and soft tissue sarcomas**
24. **Skin**
25. **Breast**

11:00-11:15 **Break**

11:15-12:00 **Abdominal Tram video**

12:30-1:00 **Breast reconstruction and implications for exercise**

1:00-2:00 **Lunch**

2:00-2:45 **Workshop – conducting a postural assessment**

2:45-3:45 **Workshop – using a goniometer**

3:45-4:00 **Break**

4:00-5:00 **Workshop - taking girth measurements**

Day Two

9:00 – 10:15	**Cancer treatment and implications for exercise** 1. Radiation 2. Chemotherapy 3. Cryosurgery 4. Photodynamic therapy 5. Testosterone ablation therapy 6. Ovarian ablation therapy 7. Hyperthermia 8. Autologous/Allogeneic BMT- Stem cell transplant 9. Immunotherapy
10:15-10:30	**Cancer treatment and weight management**
10:30-11:00	**Identification and prevention of lymphedema**
11:00-11:30	**Exercise recommendations for lymphedema patients**
11:30-12:00	**Cancer related pain**
12:00-12:30	**Mental and physical fatigue during cancer treatment**
12:30-1:30	**Lunch**
1:30- 2:00	**Benefits of exercise**
2:00-2:45	**Workshop – manual stretching techniques**
2:45-3:00	**Break**
3:00-3:45	**Working with the medical professionals/marketing & promotion strategies**
3:45-4:00	**Emergency procedures**
4:00-5:00	**Case studies/ Q&A**

Contributors

A special thanks to:

Dr. Glenn B. Gero – Dr. Gero is a board-certified doctor of naturopathy, a registered nutritional counselor, master herbalist, certified biofeedback therapist, certified exercise specialist and lifestyle coach. He is the director of Holistic Naturopathic Center in Clifton, New Jersey, and the president of the New Jersey Natural Health Professionals. For additional information, refer to: www.holisticnaturopath.com.

Dr. Joseph Camp - Dr. Camp has been in the health and fitness industry for over 25 years. After managing fitness centers in his early career, he went on to complete certifications as medical technician and fitness trainer and obtained a degree of chiropractic from Palmer College-West in 1995. He is cofounder and Club Director of Lift Fitness in Mountain View, California

Dr. Keri Winters-Stone - Dr. Winters-Stone holds Associate Professor Appointments at both Oregon Health and Science University (OHSU) and Oregon State University. Dr. Winters-Stone regularly presents her research at national and regional scientific meetings and is the author of "Action Plan for Osteoporosis". Her most recent research focuses on the effects of cancer treatment on fracture risk and the ability of exercise to improve health in cancer survivors.

Linda Matchtelinckx – Yoga instructor and international yoga educator.

Cancer Classifications

What is Cancer?

The term "cancer" refers to a group of diseases in which abnormal (malignant) cells divide and form additional abnormal cells without any order or control. In normal tissues, the rates of new cell growth and old cell death are kept in balance. In cancer this balance is disrupted. This disruption can result from uncontrolled cell growth or loss of a cells' ability to undergo "apoptosis." "Apoptosis," or "cell suicide," is the mechanism by which old or damaged cells normally self destruct. The problem with these malignant cells is that they are unable to perform the functions that they were designed for – such as to replace worn-out cells or repair damaged cells – and they continue to grow and multiply without constraint. The normal cells do not respond appropriately to the body's signals to divide only when needed and to stop when the need is fulfilled. In other words, these cells can be thought of as taking on a life of their own. The gradual increase in the number of growing cells creates a growing mass of tissue called a "tumor," or "neoplasm." If the rate of cell division is relatively rapid, and no "suicide" signals are in place to trigger cell death, the tumor will grow quickly in size; if the cells divide more slowly, tumor growth with be slower. Regardless of the growth rate, tumors ultimately increase in size because new cells are being produced in greater numbers than needed. The cells can invade and destroy healthy tissue and can spread and grow in other areas of the body through two mechanisms: invasion and metastasis. Invasion refers to the direct migration and penetration by cancer cells into neighboring tissues. Metastasis refers to the ability of cancer cells to penetrate into lymphatic and blood vessels, circulate through the bloodstream, and then invade normal tissues elsewhere in the body.

Deviation from Normal Cell Growth

Cancer tissue has a distinctive appearance when viewed under a microscope. Pathologists will look for a large number of dividing cells, variation in nuclear size and shape, variation in cell size and shape, loss of normal tissue organization, and a poorly defined tumor boundary. Sometimes pathologists will detect a condition known as "hyperplasia." This refers to tissue growth based on an excessive rate of cell division, leading to a larger than usual number of cells. Everything else in the cells' structure seems to remain normal and potentially reversible. Hyperplasia can be a normal tissue response to an irritating stimulus, for example a callus that forms on your hand when you begin playing tennis on a regular basis. Another non-cancerous condition is called "dysplasia." This, too, is an abnormal type of cell proliferation characterized by loss of normal tissue arrangement and cell structure. Often times these cells will revert back to normal behavior, but occasionally, they gradually become malignant. These areas are usually closely monitored by a professional in case they need treatment. The most severe cases of dysplasia are sometimes referred to as "carcinoma in situ." This term refers to an uncontrolled growth of cells that remains in its original location. It does, however, have the potential to develop into an invasive malignancy and, is therefore, usually removed surgically when possible. Lastly, there is invasive cancer. Unlike carcinoma in situ, this cancer has spread beyond its' original location and has begun to infiltrate into other, previously healthy, tissue. These tumors tend to grow more quickly, spread to other organs more frequently, and be less responsive to therapy. These cancers are surgically removed when possible and often accompanied by radiation and/or chemotherapy to kill any cancerous cells that have spread outside of the tumor.

What Causes Cancer?

Cancer is often perceived as a disease that strikes for no apparent reason because there are many unproven theories. Scientists don't know all of the reasons, however, many have been identified. Besides heredity, which only accounts for approximately 10% of all cancer cases, scientific studies point to three main categories of factors that contribute the development of cancer: chemicals (e.g., from smoking, diet, inhalation…), radiation (e.g., x-rays, ultraviolet, radioactive chemicals…), and viruses or bacteria (e.g., Human Papillomavirus, Epstein Barr Virus, hepatitis B…) Chemicals and radiation that are capable of causing cancer are known as "carcinogens." Carcinogens initiate a series of genetic alterations or mutations and encourage cell proliferation. This usually doesn't happen overnight. Sometimes several decades can pass between exposure to a carcinogen and the onset of cancer. Since exposure to carcinogens is responsible for triggering most cancers, we can reduce our risk by taking steps to avoid such agents whenever possible. The use of tobacco products has been implicated in one out of every three cancer deaths. To spite the Surgeon Generals' repeated warnings as well as the fact that smoking is the largest single cause of death from cancer, the tobacco industry continues to thrive. Avoiding tobacco products, cigarettes, cigars, and chewing tobacco, is the single most effective lifestyle decision you can make in an effort to prevent cancer. Although it is usually not life-threatening, skin cancer caused by exposure to sunlight is the most frequently observed type of cancer. Most of us don't take skin cancer very seriously because it is often easy to cure. Melanoma, a more serious form of skin cancer also associated with sun exposure, is potentially lethal. Once again, we choose to ignore the repeated and ever-present warnings to stay out of the sun, and continue to bask in the suns' glory for hours on end. Risk of skin cancer can be greatly reduced by wearing clothing to shield the skin from ultraviolet radiation, wearing protective sunscreen, or by avoiding direct sun exposure altogether. Actions can also be taken to avoid exposure to some of the viruses that are associated with cancers. The most common of which is the human papillomavirus (HPV), which is involved in the transmission of cervical cancer. "Safe sex," including limiting exposure to multiple sex partners, is the best way to prevent this virus which is sexually transmitted. Many carcinogens have become "occupational" hazards to those who come in contact with them on a regular basis. These include arsenic, asbestos, benzene, chromium, leather dust, naphthylamine, radon, soots, tars, oils, vinyl chloride, and wood dust. Workers who are exposed to these chemicals have a higher incidence of cancer. Although a persons' chance of developing cancer at some point in his/her lifetime is almost twice as great today as it was fifty years ago, cancer is still not considered an epidemic. The increase in identifiable cancer cases is due largely in part to increased life span because cancer is more prevalent among older people.

Obesity (being extremely overweight) raises the risk of type II diabetes, high blood pressure, heart disease, and cancer. There are approximately 40,000 cancer diagnoses in the U.S. each year are caused by obesity. In addition, being overweight and obesity cause 15% to 20% of all cancer-related deaths each year. Several studies have explored why being overweight or obese may increase cancer risk and growth. People who are obese have more fat tissue, which can produce hormones, such as insulin or estrogen, and may cause cancer cells to grow. How much a person weighs throughout various points in his or her life may also affect the risk for cancer. Research has shown that the following are modestly associated with an increased risk:

- High birth weight
- Weight gain during adulthood
- Gaining and losing weight repeatedly

Cancer cells come in all different shapes and sizes, and are classified by their aggressiveness. Cancer cells that essentially resemble their non-cancerous counterparts and can still perform some of their normal functions are described as *well differentiated.* On the flip side, the cells that are identified by their disorganized structure and their ability to divide rapidly and chaotically are known as *poorly differentiated cells.* A tumor that remains confined to its' original, or *primary* location, is referred to as *localized.* There are two ways that a cancer can spread; it can grow straight through the primary organ and directly into adjacent tissue (referred to as a *local extension or regional disease*), or in *metastatic cancer,* a colony of malignant cells can break away and ride the circulatory system to nearby lymph nodes or a distant organ where it forms a *secondary cancer*.. Sometimes, despite batteries of tests, a metastatic tumor is diagnosed, but no primary tumor is found. When this happens, the cancer is declared a *cancer of unknown primary origin.*

Classifications of Cancer

Carcinomas: make up 90% of all cancers. They arise in the epithelium, the membranous tissue that forms the inner lining and outer covering of organs, glands, and vessels, as well as the surface layer of the skin.

The following five categories make up the remaining 10%:

- **Sarcomas**- bone, cartilage, muscle, and vessels
- **Leukemias**-blood, blood cells, and bone marrow
- **Lymphomas**-lymph nodes and lymphatic system
- **Melanomas**-skin cells that produce pigment responsible for skin color
- **Gliomas**-brain and spinal cord

Types of Cancer and Their Treatments

Cancer of the Prostate

More than 232,000 cases of prostate cancer are diagnosed and treated annually in the United States, and close to 30,000 men die each year of the disease. Most men over the age of 50 will have some experience with prostate disease -- with either an enlarged prostate or cancer. African American men have the highest prostate cancer incidence in the world. Another important risk factor is a positive family history. If a man has a father or brother with the disease, his risk for developing it is twice that of a man with no family history. International studies suggest that dietary fat also may play a role in the development of the cancer. Cancer of the prostate is the most commonly seen cancer in the United States and is referred to as an *indolent* cancer, one that grows extremely slowly, sometimes taking as long as two to four years to double in size. Because the median age for diagnosis is 72, many men will elect to forgo aggressive treatment and opt for "watchful waiting" instead.

When a man enters his fifties, or thereabouts, the testicles begin to suddenly secrete testosterone, the male sex hormone. This causes the prostate to grow, increasing in size by half nearly every ten years. More than 50% of men between the ages of 60-70 suffer from a non-cancerous condition called *benign prostatic hyperplasia.* As the prostate enlarges, it compresses the urethra and the channel for the urine to pass through. Men typically feel an urgency to go to the bathroom, but are unable to void. Needless to say, the obstruction can make urination very painful. Should a cell in the prostate turn into cancer, the testosterone will spur the tumors growth like gasoline fueling a fire. Two options for stopping the supply of testosterone exist; surgically removing the testicles or by administering hormones that either halt testosterone production or block its effect.

Symptoms of cancer of the prostate include a weak or interrupted flow of urine, need to urinate frequently (especially at night), blood in the urine, inability to urinate, or difficulty starting to urinate, urine flow that is not easily stopped, painful or burning urination with radiating pain in the back, pelvis, or hips. These symptoms may indicate other prostate problems, however, they should not be ignored and the patient should be seen by a doctor to determine whether it is a cancerous or non-cancerous enlargement. Regular digital exams are recommended to detect early prostate cancer because it often does not cause any symptoms.

Treatment options

Minimally invasive surgery: is defined as a surgical procedure performed through small incisions -- usually made in the abdominal wall -- the result of which is the least possible damage to organs and surrounding tissue. The general advantages of minimally invasive surgery for patients are minimal blood loss, quicker recovery, and a better cosmetic result. The main goals of laparoscopic radical prostatectomy are to cure the patient and preserve his quality of life: both in the short term -- easier recovery after the operation -- and in the long term -- preservation of continence (ability to control urination) and potency (ability to have an erection). The surgeon begins the laparoscopic radical prostatectomy by making one incision (one centimeter, or less than half an inch, in length) around the navel in order to insert a thin, lighted tube with a telescopic camera on its tip (called a laparoscope) into the body. The camera projects an extremely clear, highly magnified visualization of the surgical area onto a screen in the operating room, by which the surgical team operates. A harmless gas is introduced into the abdomen to create a space large enough to perform the surgery. The operation is performed with specialized surgical instruments inserted through four tiny incisions in the pelvic area, and the prostate (and, if necessary, lymph nodes and surrounding tissue) is removed.

Although it is not always possible due to the size and location of the cancer, one of the primary goals of radical prostatectomy is to be "nerve-sparing." This means that the surgeon preserves the web of tiny nerves that control erection and keeps them intact. This extremely delicate and precise technique is made possible with the laparoscopic approach because of the quality of the visualization of the surgical field, due to the magnification of the surgical area and reduced bleeding

***Advantages of a Minimally Invasive Approach*:**

- Incisions are usually made in the abdominal wall – the result of which is the least possible damage to organs and surrounding tissue.
- Less blood loss during surgery
- Less pain following the operation
- Shorter recovery period
- Faster hospital discharge (65 percent of patients are discharged the day after surgery, and 30 percent two days after surgery)
- Quicker return to normal activities and work (usually within three weeks)
- Better cosmetic result -- four or five tiny incisions versus an eight-inch (or larger) incision from open surgery

Another important advantage of the minimally invasive approach is that for 90 percent of patients the Foley catheter (a thin tube inserted into the bladder to drain urine) can be removed within one week. With open surgery, the catheter usually stays in for two or three weeks following the procedure.

Radical prostatectomy: surgical removal of the prostate and surrounding tissue. The incision is made either through the lower abdomen (retropubic prostatectomy), or in a perineal prostatectomy, through the perenium, the area between the scrotum and the anus. Should a tumor extend through the prostates' fibrous capsule to infiltrate neighboring lymph nodes or scatter to distant sites, prostatectomy is no longer a viable option. An advantage to the retropubic approach is that it enables the surgeon to biopsy the nodes in the pelvic area. If no evidence of nodal involvement is found, the prostate is removed right then and there. The perineal route doesn't allow access to lymph nodes. To accomplish this, the surgeon must make one or more incisions in the abdomen and sample lymph nodes conventionally or through a laparoscope.

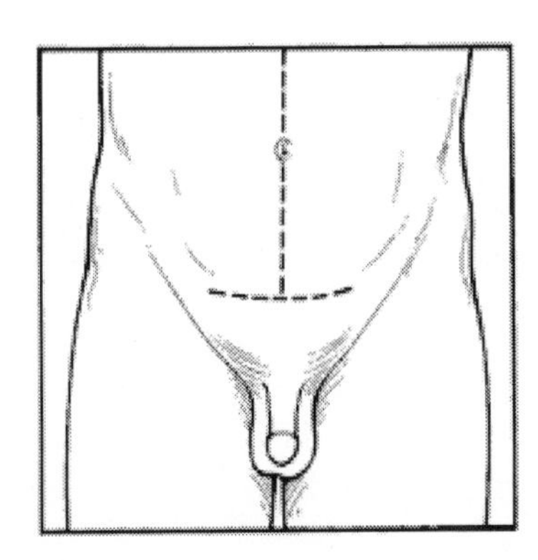

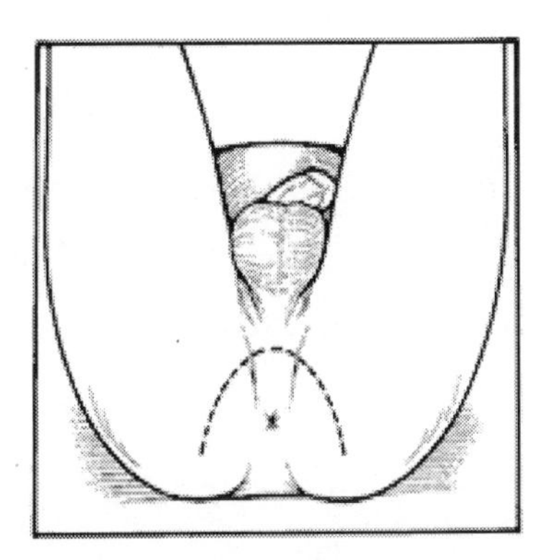

Potential side effects of radical prostatectomy:

- 65-90% impotence rate
- Bladder spasms
- Heart attack
- Stroke
- Blood clots in the legs
- 1 in 20 men are left without urinary control
- 1 in 5 men must contend with long-term stress incontinence – a cough, sneeze, laugh, or sudden movement may induce minor wetting
- Virtually all patients can expect to be incontinent for 3-4 months after surgery
- Infertility

Pelvic lymph node dissection – pelvic lymphadenectomy: surgical biopsy of the lymph nodes in the pelvic area performed through the retropubic approach.

Potential side effects of pelvic lymph node dissection:

- Lower extremity lymphedema
- Infection

Transurethral resection of the prostate (TURP): surgical removal of the prostatic tissue to relieve symptoms. The surgeon passes a flexible *cytoscope* through the urethra and into the prostate. An electrified wire loop is inserted through the scope and used to cut away the tumor or nonmalignant tissue that is obstructing the flow of urine. This procedure would typically be used for a patient who was diagnosed late in the course of the disease and didn't respond to traditional medical therapy, or who chose watchful waiting and then had progressive local disease that caused bladder-outlet obstruction. It is a palliative treatment for relieving pain and restoring normal urine flow. The operation usually takes about an hour. After surgery, a catheter is inserted through the penis into the bladder. It remains in place for 1 to 3 days to help urine drain while the prostate heals. The patient can usually leave the hospital after 1 to 2 days and return to work in 1 to 2 weeks.

Potential side effect of TURP:

- 1 in 10 men will experience partial impotence or incontinence
- Recurring urinary tract infections
- 9 in 10 men will experience *retrograde ejaculation*, in which the semen discharge into the bladder instead of through the urethra and out the penis
- Difficulty urinating – may last several months
- Bloody urine after surgery

Bi-lateral orchiectomy – is the surgical removal of the testicles, through a small incision in front of the scrotum. This is considered the most effective method of hormonal ablation therapy.

Potential side effects of bi-lateral orchiectomy:

- Hot flashes or flushing
- Decreased sexual desire

- Fatigue
- Weight gain
- Potential impotence
- Breast tenderness or growth
- Osteoporosis and reduction of muscle tone over prolonged periods
- Lower extremity lymphedema

External beam radiation – is a choice for men who want to avoid undergoing an operation or for those whose health rules out major surgery. As a treatment for stage I and II prostate tumors, it yields the same 10-year outcome as radical prostatectomy, but with fewer complications.

Potential side effects of external beam radiation:

- Fatigue
- Skin reactions
- Frequent, urgent urination
- Difficult urination
- High-energy rays may irritate the rectum (*radiation proctitis*) causing painful bowel movements, burning, bleeding, hemorrhoids, cystitis, and diarrhea
- Abdominal cramping
- Hemorrhoids
- Cystitis
- 30-50% of patients will become impotent
- Only 5% of patients will become incontinent

Three dimensional conformal radiation – uses a CT scanner and computer to conform the radiation field precisely to the shape of the prostate and give a higher dose than conventional external-beam radiation therapy. This type of therapy not only causes fewer symptoms, but those that do arise are significantly milder. This method of delivery provides incredible results in terms of prolonging remission.

Seed-implant therapy (brachytherapy) – is considered internal radiation. The patient must be unconscious or anesthetized from the waist down while the doctor inserts a needle containing a tiny radioactive pellet through the perineum into the prostate, where it will emit a continuous dose of radiation for several days before eventually becoming inert. During the course of treatment anywhere from several dozen to more than a hundred seeds are implanted. The seeds are small (about the size of sesame seeds) and cause little or no discomfort. This treatment works best for small tumors. Because it doesn't affect the surrounding tissue, side effects such as incontinence and impotency are greatly diminished. For patients with more advanced stages or aggressive disease the combination of a permanent seed implant followed two months later by a five- to six-week course of highly focused, intensity-modulated radiotherapy is another effective means of safely delivering a high dose of radiation to the cancer. High-Dose-Rate Brachytherapy is used for patients whose cancer is somewhat more advanced or aggressive. High-dose-rate (HDR) or temporary brachytherapy is used to deliver weeks' worth of radiation in three to four minutes. In this procedure, radiotherapists send seeds into and out of the prostate on three occasions over the course of a day and a half.

Potential side effects of brachytherapy:

- Impotence
- Urinary urgency and frequency
- Occasional rectal discomfort or bleeding

Hormonal ablation therapy (medical castration) – when the cancer reaches stage IV, it has usually spread to the lymph nodes, or to the bone, liver, and/or lung. Hormonal therapy is not a cure, but can minimize the cancer's spread, often for a period of years. It may also be used palliatively to shrink a tumor that is causing pain or encroaching on the urethra. The purpose of hormonal ablation therapy is to stop the testicles' production of testosterone, which causes the prostate-tumor cells to grow. Because it is systemic, it can affect cancer cells throughout the body.

Types of drugs (see page 181 for side-effects and warnings)

LHRH agonists – work to block production of *luteinizing hormone-releasing hormone. LHRH* triggers a gland at the base of the brain to secrete *luteinizing hormone*, which in turn stimulates testosterone production. Even though LHRH agonists cost more and require more frequent doctor visits, most men choose this method over medical castration. When LHRH analogs are first given, testosterone production increases briefly before falling to very low levels. This is called *"flare"* and it results from the complex way in which LHRH agonists work.

Men whose cancer has spread to the bones may experience bone pain. If the cancer has spread to the spine, even a short-term increase in growth could compress the spinal cord and cause pain or paralysis. Anti-androgens, if given for a few weeks when starting treatment with LHRH agonists, can help prevent "flare."

Diethylstilbestrol (die-eth-el-still-bess-troll), a*lso Known As: Stilboestrol.* It is given orally (by mouth). It is a synthetic estrogen that functions like the LHRH agonists. It was the first chemical alternative to an orchiectomy.

Antiandrogen medication – a small amount of male hormone is made by the adrenal glands, and may not be affected by bilateral orchiectomy or LHRH agonists. An antiandrogen is a medication that can block the effect of the remaining male hormone on prostate cancer cells. Antiandrogens are pills often given to patients in addition to orchiectomy or LHRH agonists.

Bone-protecting treatments - the most common site of distant spread of prostate cancer is the bones. Most symptoms of advanced prostate cancer are caused by the presence of disease in the bone. These symptoms can be mitigated with a drug called Zoledronic acid, which can slow the spread of disease, reduce the development of bone pain, inhibit bone fractures, and confer other beneficial effects. Zoledronic acid is most commonly given to patients whose cancer is no longer responding to hormones, but it may also be given to prevent the bone thinning and weakening that results from hormonal treatments.

Cryosurgery – uses cold-tipped probes to destroy cancerous prostate tissue. A man who has a small localized tumor that did not respond to radiation therapy and who could not withstand other treatments would be eligible for this procedure because it is minimally invasive. Because it is not as invasive as other procedures, there is less pain, bleeding, and other complications. Unfortunately, it is rarely covered by insurance because the effectiveness of this technique is still being assessed, and very few medical centers offer this type of treatment.

Potential side effects of cryosurgery:

- Urinary incontinence (often temporary)
- Impotence (often temporary)
- Itching or burning with urination
- Blood in the urine
- Scrotal edema

Chemotherapy – is usually reserved for patients whose prostate cancer has spread outside of the prostate gland and for whom hormone therapy has failed. Chemotherapy for prostate cancer has been shown to prolong patients' lives, reduce pain from bone metastases, and enhance overall quality of life. The benefits of chemotherapy for advanced prostate cancer are comparable to the benefits seen in patients with other advanced solid tumors who receive chemotherapy. The agent now considered the standard of care for prostate cancer patients is docetaxel (Taxotere®), which is directed at the structural skeleton of the cell.

Cancer of the Lung

Lung cancer is by far the leading cause of death from cancer. Tobacco use is to blame for 85% of all lung cancers. The other 15% arise mainly from occupational and environmental exposures to radon, asbestos and second-hand cigarette smoke. Lung cancer has replaced breast cancer as the prime cause of cancer death for women. The incidence of lung cancer decreases when smoking is stopped; after about fifteen years the risk is the same as that of non-smokers. Nearly half of all lung cancer patients will be diagnosed with metastatic disease. Lung cancer cells that break off from the original tumor often form a secondary tumor in the brain. An isolated secondary tumor is potentially curable if treated properly with chemotherapy and surgery. Symptoms of lung cancer often depend on the tumor's location. A persistent cough is the most common symptom. Additionally, wheezing, shortness of breath, constant chest pain, spitting up phlegm (especially in the early morning), streaks of blood coughed up in the phlegm, hoarseness, fever, neck enlargement, repeated pneumonia or bronchitis, weight loss, and arm and shoulder pain may all be symptoms of lung cancer.

Types of lung cancer

- **Small cell carcinoma** – is also referred to as oat-cell cancer because the tiny cells, when viewed under a microscope, actually look like oats. The tumor will typically appear in the central portion of the lung and is extremely virulent, spreading to the lymph nodes and other organs extremely quickly. Without treatment, small-cell carcinoma is usually fatal within a month or two.

- **Non-small-cell lung cancer** – Three types fall under this category:
    - ***Squamous-cell carcinoma (epidermoid)*** usually takes root in the bronchi. It spreads slower than any other type of lung cancer. It is most common in men and account for about 20% of all lung cancers.

- *Adenocarcinoma* is the most common form of lung cancer in people who have never smoked, particularly women. It typically begins in the outer perimeter of the lungs and under the bronchial lining.

- *Large-cell carcinomas* are named for their large, abnormal-looking cells. They are a group of cancers that are also found along the lung's outer border.

- **Mesothelioma -** is a relatively rare cancer that affects the membrane lining the chest or abdominal cavity. There may be a connection between asbestos exposure and mesothelioma.

Treatment options:

Prophylactic cranial radiation – is a palliative form of radiation that may be administered to the brain of small cell lung cancer patients to prevent tumors from forming on the brain.

Mediastinoscopy: is a relatively non-invasive procedure done under general anesthesia; a small incision is made in the lower neck and a special instrument is inserted along the windpipe to take biopsies from the enlarged nodes.

Bronchoscopy: this procedure is carried out for diagnostic purposes or for treatment. A rigid or flexible instrument is inserted into the airway and allows the physician to see the voice box, windpipe and bronchi. A flexible bronchoscope is often used after topical anesthesia and can be inserted orally or via a nostril. Rigid instruments are almost always used after general anesthesia.

Endobronchial Stenting: inserts a plastic tube into the airway as a palliative treatment for breathlessness.

Wedge resection / segmental resection: surgical removal of a small portion of the lung. Usually performed in situations where the doctor is looking to relieve symptoms rather than effect a cure.

Contraindications for surgery

- A heart attack in the preceding three months
- A tumor in the opposite lung
- Malignant *pleural effusion,* a pooling of cancer-contaminated fluid in the pleural space between the membranes that encase each lung
- Dissemination of the cancer to sites outside of the chest

Thoracotomy: an incision is made on the side of the chest between the ribs.The ribs are spread apart so that the surgeon can see the chest cavity. A small piece of rib may also be removed so that the surgeon is able to get out all of the cancer.

Lobectomy: the standard minimal surgery for lung tumors, in which a section (lobe) is taken out. The left lung has two lobes, the right lung three. A *bilobectomy* removes two lobes of the right lung.

Pneumonectomy: surgical removal of the entire lung on either the left or right side.

Sleeve pneumonectomy: surgical removal of the entire lung as well as lower trachea. The airway must be reconstructed following this procedure.

Extended resection: reconstructive procedure after part of chest wall, left atrium, and diaphragm are removed.

Sternotomy: doctor may split the sternum through it's midline in order to see both sides of chest to locate undetected cancer. In some cases the doctor will also remove a small portion of the lung.

***Video-assisted thoracoscopic surgery (VATS)*:** may be done before or instead of a thoracotomy. The procedure involves inserting a long, thin tube (videoscope) with a camera attached and small surgical instruments into the chest through small incisions made between the ribs. The VATS method may be used to:

- Confirm the diagnosis of lung cancer
- Biopsy lymph nodes
- Perform a wedge resection to remove the cancer and the lung tissue surrounding the cancer
- Remove the segment (lobe) of the lung that contains the cancer

Potential side effects for all lung surgeries:

- May need a respiratory therapist to learn special breathing exercises and techniques for coughing effectively so that secretions from the lung don't accumulate in the air passages.
- The recovery from any lung surgery averages six weeks, although one in ten patients continue to feel pain in the incision site longer.

- Because surgeons must cut between ribs to get to the lungs, clients may experience pain in their ribs until they fully heal.
- Occasionally, part of a rib is removed during surgery, and a rib fracture is not uncommon
- Soreness in the chest and arm
- Shortness of breath

Chemotherapy – uses anticancer agents to systemically destroy cancerous cells. It may also be used to shrink tumors, either to reduce pain, or to make surgery easier. Chemotherapy is a systemic treatment that travels through the bloodstream and kills cancerous cells anywhere in the body by interfering with cell growth and division. Chemotherapy is the treatment of choice for small cell lung cancer which is not controllable with surgery. It may be given in conjunction with radiation therapy, before or after surgery, if the tumor is localized, but needs to be shrunk before it can be removed. It may also be used for non-small cell lung cancers in patients who are not candidates for surgery or radiation therapy.

Potential side effects of chemotherapy:

- Hair loss
- Dry skin or rash
- Weight gain/loss
- Fatigue
- Nausea, vomiting, or diarrhea
- Decreased appetite
- Nerve damage causing arm or leg tingling and numbness
- Muscular weakness / fatigue
- Immunocompromization
- Damage to the circulatory system and heart muscle (cardiomyopathy or accelerated atherosclerosis)
- Diabetes and osteoporosis
- Peripheral neuropathy of the feet and hands

External beam radiation – radiation uses high-dose x-rays to shrink tumors and kill cancer cells that may have been left behind after spreading away from a tumor that was removed by surgery. These remaining cells are often referred to as *hidden cells*, because they are not detected at the time of surgery. The procedure kills cancerous cells by permanently damaging their DNA in a way that causes them to lose their ability to function and then die. This procedure will inevitably rob patients permanently of some degree of lung function.

Potential side effects of external beam radiation:

- Fatigue
- Dry, sore throat
- Skin irritation
- Coughing with or without shortness of breath
- Lung scarring causing shortness of breath

Photodynamic therapy – entails injecting a photosensitizing agent into the circulation. Ironically it does not treat cancer systemically, but remains local. Although the drug enters all of the body's cells, it clears from non cancerous cells rapidly. Forty-eight hours after the injection, a red beam of argon laser light is sent through a fiber optic scope placed against the tumor, setting off a chemical reaction that destroys the cancer cells. This treatment may also be used palliatively for the treatment of tumors that are obstructing the esophagus and interfering with swallowing. For cancers of the skin, the drug is usually applied to the skin as a cream. The area is then covered with a dressing to protect it and the patient will have to wait for about 3-6 hours (to allow time for the cream to work) before they have their treatment. The length of time they wait varies depending on the skin condition being treated.

When they come back for treatment they will be asked to sit or lie down in a comfortable position. A strong light is then shone directly onto the affected skin. This can take 8-45 minutes depending on the particular procedure they're having. The treatment is often repeated a week later.

An advantage of PDT is that it causes minimal damage to healthy tissue. However, because the laser light currently in use cannot pass through more than about 3 centimeters of tissue, PDT is mainly used to treat tumors on or just under the skin or on the lining of internal organs. PDT is only used for small, early cancers that are inoperable due to medical reasons or for small, multiple lung cancers. It is not an effective form of treatment for tumors that extend through the bronchial wall or have metastasized to regional lymph nodes.

Potential side effects of PDT:

- Photodynamic therapy makes the skin and eyes sensitive to light for 6 weeks or more after treatment.
- Patients are advised to avoid direct sunlight and bright indoor light for at least 6 weeks.
- If patients must go outdoors, they need to wear protective clothing, including sunglasses.
- Itching, stinging, or burning of lesions
- Hypo/hyper pigmentation
- Edema
- Shortness of breath
- Coughing up blood
- Fever
- Pneumonia
- Bronchitis

Stereotactic radiosurgery – using the same computerized three-dimensional guidance as stereotactic surgery, one of several types of machines sends multiple narrow beams of high-dose radiation penetrating the tumor from different angles. Typically most lesions will stop growing soon after treatment, but several weeks or months may pass before it becomes apparent. The tumor will gradually shrink and disappear. Many patients with recurrent non-small cell lung cancer are eligible for clinical trials. Radiation therapy may provide excellent palliation of symptoms from a localized tumor mass.

Internal radiation therapy – most lung tumors grow in the bronchial tubes that deliver oxygen and take away carbon dioxide. The obstruction of an airway can give rise to *postobstructive pneumonia,* in which secretions build up, making it difficult for patients to catch their breath. Until this dangerous condition is brought under control, chemotherapy can not be administered. Internal radiation therapy is also known as *endobronchial radiation,* delivers radiation directly to the tumor using a thin catheter that is fed down the throat and windpipe. The dose of radiation is delivered continually over several days, while minimizing some of the side effects of the surrounding tissue. This procedure is highly effective for relieving pneumonia, which then allows for chemotherapy. The downfall is that it leaves patients radioactive for a few days. The *radionuclides* are sealed in containers such as pellets, needles, capsules, or wires, which are implanted under general anesthesia.

Endobronchial stenting – is used to open up a blocked airway endoscopically. The doctor burns out (*fulgurates*) a portion of the tumor. While the patient is unconscious, a cylinder-like stent is passed through a bronchoscope and placed in position, creating a tunnel through the obstruction.

Cancer of the Colon or Rectum

Colorectal cancer refers to cancer of the large intestine. The first five feet of the intestine is called the colon (bowel) and the last six to eight inches is called the rectum. It is sometimes difficult to pinpoint a tumor in the large bowel and, therefore, the colon and rectum are often grouped together as the *colorectum.* Ninety-nine percent of all colorectal cancers begin with *adenomatous polyps,* small mushroom-shaped masses that form on the mucosa (inner lining) of the colon, and sometimes the rectum as well. It is not uncommon to find polyps in men and women over the age of 65, however, they are usually benign and nothing to be alarmed about. Unfortunately there is only one way to determine if a growth is cancerous or not, remove it surgically. The most common warning signs for colon cancer are changes in bowel habits, such as constipation or diarrhea, or changes such as persistent narrowing of the size of the stools. Additionally, gas pains, cramps, or bleeding that last for more than two weeks.

Treatment options:

Polypectomy: a technique for removing colorectal polyps without having to cut into the abdomen. A flexible fiber-optic scope is inserted into the rectum and carefully maneuvered throughout the colon. A wire loop is then passed through the instrument and over the tip of the polyp (*stalk*). By generating a painless electrical current, the polyp is cut and the mass is eliminated. If the polyp is too large to be removed in this manner, surgery will be required.

Sharp mesorectal incision: is a procedure for cancers that have grown through the wall of the rectum or involve the lymph nodes which may require more extensive surgery. This approach allows the delicate removal of all cancerous tissue in and around the rectum, but carefully avoids severing the nerves that are involved in sexual and urinary function, and also allows most patients to avoid a permanent colostomy. In some patients, such as men with large prostate glands, such techniques may not be feasible, and "coloanal reconstruction" is needed. This approach allows the surgeon to remove the rectum, but avoids the need for a permanent colostomy by sewing the upper colon directly to the anus with the use of specialized equipment.

Colon resection (Colectomy): is the most common operation for colon cancer. The surgeon removes the cancer and surrounding normal colon. The *mesentery* (fatty connective tissue that holds the colon in place) is also removed as well as ten to thirty adjacent lymph nodes. If the doctor is not able to sew the 2 ends of the colon back together, a stoma (an opening) is made on the outside of the body for waste to pass through. This procedure is called a colostomy. Sometimes the colostomy is needed only until the lower colon has healed, and then it can be reversed. If the doctor needs to remove the entire lower colon, however, the colostomy may be permanent. The day before surgery, the patient will probably be told to completely empty their bowel. This is done with a bowel preparation, which may consist of laxatives and enemas. After surgery there will probably be some pain that will require pain medicines for 2 or 3 days. For the first couple of days intravenous (IV) fluids will be administered. During this time the patient may not be able to eat or may be allowed limited liquids, as the colon needs some time to recover. But a colon resection rarely causes any major problems with digestive functions, and they should be able to eat solid food again in a few days.

Potential side effects of colectomy:

- Lower extremity lymphedema
- Temporary constipation or diarrhea
- Irritation of the skin around the stoma

Electrofulgeration: is a type of electrosurgery that destroys cancerous tissue with heat generated by a high frequency current.

Local excision – a small tube is placed in the colon or rectum, by way of the anus, and the cancer is cut out.

Bowel resection (*proctectomy*) – is the surgical removal of the large intestine or, more often, a portion of it, with the severed ends being sutured together. Area lymph nodes are also dissected and sent to pathology to be analyzed. If the bowel must be excised, or the healthy tissue cannot be reconnected, an artificial opening, or *stoma*, for eliminating solid waste must be surgically created in the lower abdomen. This procedure is called a *colostomy.*

Potential side effects of proctectomy:

- Lower extremity lymphedema
- Temporary constipation or diarrhea
- Irritation of the skin around the stoma

Pelvic exenteration – the surgeon removes the rectum and nearby organs such as the bladder, prostate (in men), or uterus (in women) if the cancer has spread to these organs. The patient will need a colostomy after pelvic exenteration. If the bladder is removed, they will also need a urostomy (opening where urine exits the front of the abdomen and is held in a portable pouch

Abdominal-perineal resection – is more involved than a low anterior resection. It can be used to treat some stage I cancers and many stage II or III rectal cancers in the lower third of the rectum (the part nearest to the anus), especially if the cancer is growing into the sphincter muscle (the muscle that keeps the anus closed and prevents stool leakage). Here, the surgeon makes one incision in the abdomen, and another in the perineal area around the anus. This incision allows the surgeon to remove the anus and the tissues surrounding it, including the sphincter muscle. Because the anus is removed, the patient will need a permanent colostomy to allow stool a path out of the body. As with a low anterior resection, the usual hospital stay for an Abdominal-perineal resection is 4 to 7 days, depending on the overall health of the patient. Recovery time at home may be 3 to 6 weeks

Low anterior resection (LAR) – some stage I rectal cancers and most stage II or III cancers in the upper third of the rectum (close to where it connects with the colon) can be removed by this procedure. The part of the rectum containing the tumor is removed without affecting the anus. The colon is then attached to the remaining part of the rectum so that after the surgery bowels can be moved in the usual way. The surgeon makes an incision in the abdomen and then removes the cancer and a margin of normal tissue on either side of the cancer, along with nearby lymph nodes and fatty and fibrous tissue around the rectum. The colon is then reattached to the rectum that is remaining so that a permanent colostomy is not necessary. If radiation and chemotherapy have been given before surgery, it is common for a temporary ileostomy to be made (where the last part of the small intestine -- the ileum -- is brought out through a hole in the abdominal wall). Usually this can be reversed (the intestines reconnected) about 8 weeks later.

Chemotherapy – uses anticancer agents to systemically destroy cancerous cells. It may also be used to shrink tumors, either to reduce pain, or to make surgery easier. Chemotherapy is a systemic treatment that travels through the bloodstream and kills cancerous cells anywhere in the body by interfering with cell growth and division.

Potential side effects of chemotherapy:

- Hair loss
- Dry skin or rash

- Weight gain/loss
- Fatigue
- Nausea, vomiting, or diarrhea
- Decreased appetite
- Nerve damage causing arm or leg tingling and numbness
- Muscular weakness / fatigue
- Immunocompromization
- Damage to the circulatory system and heart muscle (cardiomyopathy or accelerated atherosclerosis)
- Diabetes and osteoporosis
- Peripheral neuropathy of the feet and hands

Radiation therapy – is not used often for colon cancer due to the fact that the colon is constantly in motion, contracting and expanding as it moves waste from one end to the other. The risk of irradiating nearby organs tends to outweigh the potential benefits. By contrast, however, the rectal area is "fixed" and easily irradiated without damaging neighboring tissues. For selected patients whose rectal cancer has recurred in the pelvis after surgery and radiation therapy there is a technique called intraoperative radiation therapy (IORT). The surgeon removes the tumor, and while still in the operating room, the patient receives a high dose of radiation. Because the radiation therapy is done during the surgical procedure, and can be delivered to a precisely defined area in the pelvis, it is possible to use a higher-than-usual -- and therefore more effective -- dose of radiation.

Potential side effects of radiation therapy:

- Fatigue
- Diarrhea
- Abdominal pain and cramping
- Straining while urinating or defecating
- Rectal pain
- Rectal bleeding
- Mucus discharge from the rectum

- Fecal incontinence

Radiofrequency ablation - uses high-energy radio waves for treatment. A thin, needle-like probe is placed through the skin and into the tumor. Placement of the probe is guided by ultrasound or CT scans. An electric current is then run through the tip of the probe, causing the release of high-frequency radio waves that heat the tumor and destroy the cancer cells.

Ethanol (alcohol) ablation - is known also as *percutaneous ethanol injection (PEI).* This procedure involves injecting concentrated alcohol directly into the tumor to kill cancer cells. This is usually done though the skin using a needle, which is guided by ultrasound or CT scans.

Cryosurgery (cryotherapy) – uses cold-tipped probes filled with liquid nitrogen or carbon dioxide to destroy cancerous tissue. A patient who has a small localized tumor that did not respond to radiation therapy and who could not withstand other treatments, would be eligible for this procedure because it is minimally invasive. Because it is not as invasive as other procedures there is less pain, bleeding, and other complications.

Hepatic artery embolization - is sometimes an option for tumors that cannot be removed. This technique is used to reduce the blood flow in the hepatic artery, the artery that feeds most cancer cells in the liver. This is done by injecting materials that plug up the artery. Most of the healthy liver cells will not be affected because they get their blood supply from the portal vein. The surgeon puts a catheter into an artery in the inner thigh and threads it up into the liver. A dye is usually injected into the bloodstream at this time to allow the surgeon to monitor the path of the catheter via angiography, a special type of x-ray. Once the catheter is in place, small particles are injected into the artery to plug it up. Embolization also reduces some of the blood supply to the normal liver tissue. This may be dangerous for patients with diseases such as hepatitis and cirrhosis, who already have reduced liver function.

Cancer of the Bladder

Men are more than twice as likely to develop bladder cancer than women. Most cancers occur in the bladders walls (mucosa). Seventy to eighty percent of bladder cancers are superficial at the time of diagnosis (it hasn't penetrated the bladder's wall). Most can be treated without having to remove the bladder, and the five-year survival rate approaches 95%. However, if the cancer has invaded the bladder wall, consideration must be given to removing the bladder or beginning a combination of radiation and chemotherapy. Symptoms of bladder cancer include blood in the urine, pain with urination, and need to urinate often or urgently.

Treatment options:

Transurethral resection with fulguration: a flexible cytoscope is inserted through the urethra and into the bladder. An electrified wire loop is then passed through the scope. It is "heated up" and used to cut the tumor off of the bladder wall, simultaneously coagulating the site to prevent bleeding.

Cystectomy – is the surgical removal of the bladder. A segmental cystectomy resects only the cancerous section of the bladder. This procedure is rarely applicable because of the propensity for the tumor to grow in multiple areas of the organ. A radical cystectomy removes the bladder, the surrounding organs, and some of the pelvic lymph nodes may also be taken out. In women, this extensive operation entails taking out the reproductive organs, bringing on instant menopause. In men, the prostate is removed, often resulting in impotence.

Potential side effects of cystectomy:

- Lower extremity lymphedema
- Menopausal symptoms
- Impotence
- There is a risk of excessive bleeding, urinary tract infections, urine leakage (incontinence), and blockage of urine flow in radical cystectomies

Urostomy- the two ureter tubes that transport urine from the kidneys to the bladder are implanted into a small loop of the small intestine, which is then brought out through the abdominal wall. A soft plastic external pouch is worn on the abdomen and attaches to the protruding opening or stoma, and collects the urine.

Continent urostomy- another type of urostomy is the continent diversion. Here, the pouch created from the piece of intestine has a valve created. The valve allows the urine to be stored in the pouch and emptied several times each day by placing a drainage tube (catheter) into the hole. Some patients prefer this because there is no bag on the outside.

Potential side effects for both urostomy procedures:

- Wound infections
- Urine leaks (incontinence)
- Pouch stones
- Blockage of urine flow

Chemotherapy – uses anticancer agents to systemically destroy cancerous cells. It may also be used to shrink tumors, either to reduce pain, or to make surgery easier. Chemotherapy is a systemic treatment that travels through the bloodstream and kills cancerous cells anywhere in the body by interfering with cell growth and division.

Potential side effects of chemotherapy:

- Hair loss
- Dry skin or rash
- Weight gain/loss
- Fatigue
- Nausea, vomiting, or diarrhea
- Decreased appetite
- Nerve damage causing arm or leg tingling and numbness
- Muscular weakness / fatigue
- Immunocompromization
- Damage to the circulatory system and heart muscle (cardiomyopathy or accelerated atherosclerosis)
- Diabetes and osteoporosis
- Peripheral neuropathy of the feet and hands

Intravesical chemotherapy – once bladder cancer has occurred, even if it is superficial, the organs' entire lining is at risk for a future cancer. Therefore, after the urologist removes all of the cancer he can see through the cytoscope, the patient is asked to return for drug treatment to reduce the chance of new malignancies developing. One or more drugs are instilled directly into the bladder through a catheter tube. The tube is clamped for a period of time so that the drug remains in contact with the bladder lining.

Radiation therapy – is an option for people who are considered poor surgical risks or who are adamant about not having their bladder removed.

Potential side effects of radiation therapy:

- Frequent urination
- Difficulty starting urination
- Painful urination
- Diarrhea
- Abdominal pain
- Straining while urinating or defecating
- Decreased bladder capacity causing frequent urination

Immunotherapy – cancer has a way of fooling the immune response by taking a normal cell within the body and making it so that the immune system doesn't recognize it as foreign. Immunotherapy attempts to overcome the body's tolerance and get it to reject the cancer just as it would a transplanted organ. Under normal conditions our bodies exterminate cancer cells before they can evolve into a tumor. Usually the disease manages to take hold during times when our immunity is down.

Potential side effects of immunotherapy:

- Flu-like symptoms for a short time

Cancer of the Kidney (Renal-Cell Carcinoma)

Renal cell carcinoma accounts for eighty-five percent of adult kidney cancers. They originate in the lining of the *tubules* that run throughout the organ. A second, rare form, ***transitional-cell carcinoma***, arises in the *renal pelvis*, located in the kidney's midsection. This form of kidney cancer closely resembles bladder cancer and is treated in much the same way. About fifty percent of all patients are diagnosed with local disease, stage I and II, which is frequently curable. At stage III, surgery can still prolong life and, sometimes, bring about a cure. If a kidney cancer appears at all resectable, an attempt will be made to remove it because there is no other effective treatment.

Treatment options:

Partial nephrectomy- excises only the cancerous portion of the kidney and a margin of healthy surrounding tissue. A patient undergoing a partial nephrectomy typically has no kidney on the opposite side, or suffers from impaired renal function, so that removal of the entire kidney will make him/her dependent on dialysis. The most common incision sites are under the ribs on the same side as the cancer, the middle of the abdomen, or in the back behind the cancerous kidney. Partial nephrectomy yields comparable results to complete nephrectomy in patients with small tumors (less than 4 centimeters), while maintaining functioning kidney tissue. In ***hemodialysis***, a machine called a *dialyzer* carries out the task of filtering blood three times a week. ***Peritoneal dialysis*** does not require a machine; the process uses gravity to fill and empty the abdomen. A typical prescription requires three or four exchanges during the day and one long (usually 8 to 10 hours) overnight exchange as the patient sleeps. The dialysis solution used for the overnight exchange may have a higher concentration of dextrose so that it removes wastes and fluid for a longer time. The function is performed several times a day, but inside the person's body.

Simple nephrectomy- takes out the entire kidney.

Radical nephrectomy- is the most common procedure to treat renal cell carcinoma. The surgeon removes the kidney, the adjacent adrenal gland, the surrounding fat, and *Gerota's fascia* (a filmy sac that envelopes the kidney and its ureter tube). The most common incision sites are under the ribs on the same side as the cancer, the middle of the abdomen, or in the back behind the cancerous kidney. Five to twenty lymph nodes may be removed as well.

Laparoscopic nephrectomy – this procedure is done through several small incisions. Typically, one of the incisions is made larger n order to remove the kidney (it is still smaller than it would be with the traditional, open approach). The benefits include a shorter hospital stay, a quicker recovery time, and less pain.

Potential side effects of all nephrectomies:

- Bleeding during or after surgery that may require blood transfusions
- Infection
- Damage to the spleen, pancreas, aorta, vena cave, and large or small bowel
- Pneumothorax (unwanted air in the chest cavity)
- Incisional hernia
- Kidney failure (if remaining kidney fails to function properly)
- Lower extremity lymphedema

Immunotherapy – renal-cell carcinoma is one of the cancers that are most sensitive to immunomodulation. Although cures are very rare from immunotherapy, the responses are often dramatic. Interleukin-2 (brand name Proleukin) seems to have more promise than interferon alfa-2A (brand name Roferon-A). Interleukin has produced long-lasting remissions in five percent of the people using it. Both of these treatments cause the cancers to shrink to about half of their original size in about 10-20% of patients. The downfall is that the high doses that have been used thus far are extremely toxic and, therefore, are not used on people with overall poor health.

Potential side effects of immunotherapy:

- Low blood pressure
- Extreme fatigue
- Fluid buildup in the lungs
- Trouble breathing
- Kidney damage
- Heart attacks
- Intestinal bleeding
- Diarrhea or abdominal pain
- High fever and chills
- Rapid heart beat

- Mental changes

Radiation therapy – external beam radiotherapy may be used to relieve symptoms in patients unable to undergo surgery.

Potential side effects of radiation therapy:

- Nausea, vomiting
- Diarrhea
- Abdominal pain
- Straining while urinating or defecating

Arterial embolization - is a rarely performed procedure that injects a foreign substance into the targeted artery. It is extremely painful and has not been shown to be effective in treating kidney cancer. The one exception may be to stop life-threatening bleeding in someone with an unresectable cancer and no evidence of metastatic disease.

Hormonal therapy – because kidney cancers have been found to contain high levels of progesterone receptors, it was once thought that hormone therapy was promising. As with arterial embolization, hormonal therapy has been discounted except as a palliative measure.

Cancer of the Pancreas

Pancreatic cancer is truly "silent and deadly." It will silently host the presence of a growing cancer for some time. When symptoms finally appear, they are frequently mistaken for other disorders such as hepatitis, gallstones, and diabetes. By the time the tumor has been detected, it has usually spread outside the pancreas. Only about 10% of pancreatic cancers are contained within the pancreas. More than ninety-five percent of pancreatic cancers arise in the exocrine tissue, typically producing jaundice from an accumulation of bile in the bloodstream. Tumors in the body or tail of the pancreas are far away from the common bile duct and typically won't trigger jaundice in the early stages. The number of cases of exocrine cancer is increasing in the United States as well as other parts of the world. Though it only accounts for about three percent of all new cancer cases, it is the fifth leading cause of death from cancer. The other five percent arise in the endocrine pancreas and are referred to as ***islet-call carcinoma***. This type of tumor is less aggressive than exocrine pancreatic cancer and has a more favorable prognosis. Recurrence of cancer of the pancreas is very likely. Unless the cancer is found at a very early stage, only three percent of patients will survive more than five years after diagnosis. Symptoms of pancreatic cancer include upper abdominal or lower back pain that gradually worsens and is most severe at night, back pain that is aggravated by lying flat and relieved by sitting up or lying in a fetal position, pain in the back that is relieved by bending forward or by standing, pain that occurs several hours after meals and is more severe at night, weight loss, change in bowel habits, diarrhea or greasy stools, severe constipation, jaundice, and the sudden onset of diabetes. The type of operation will depend on the stage of the disease, age, and overall condition. After the pancreas are removed the recovery period will be a long one and the doctor will prescribe a permanent low-sugar, low-fat diet, often with the addition of vitamin K. Because of the restricted diet, it is often difficult to gain weight, which makes recovery slower than usual.

Treatment options:

Whipple procedure (pancreatoduodenectomy) - surgical removal of the head and neck of the pancreas, sometimes the body of the pancreas, the gallbladder, part of the stomach, the lower half of the bile duct, part of the small intestine, and some surrounding tissue. This is the most common procedure to remove a cancer in the exocrine pancreas. The remaining bile duct is attached to the small intestine so that the bile from the liver can continue to enter the small intestine. After this operation, the patient can generally produce adequate amounts of insulin and digestive enzymes.

Potential side effects of the whipple procedure:

- Possibility of developing diabetes or, worsening diabetes
- Leaking from the various "connections"
- Infection
- Bleeding
- Trouble with the stomach emptying itself after eating
- Stomach may be paralyzed up to 4-6 weeks after surgery – may require feeding tube
- Temporary weight loss

Distal pancreatectomy – surgical removal of the body and tail of the pancreas, sometimes part of the body, and usually the spleen. This procedure is used most often with islet cell carcinoma found in the tail and body of the pancreas. It is not used very often with exocrine cancers because these cancers have usually spread.

Potential side effects of distal pancreatectomy:

- Insulin dependence
- Immunocompromization and overwhelming post-splenectomy infections

Total pancreatectomy – surgical removal of the entire pancreas as well as the spleen, and most of the adjacent lymph nodes. This procedure is not often used to treat exocrine cancers because there doesn't seem to be any marked advantage for removing the entire pancreas. While it is possible to live without a pancreas, when the islet cells are removed, people are unable to produce insulin. They then develop diabetes and become dependent on insulin.

Potential side effects of total pancreatectomy:

- Lower extremity lymphedema
- Insulin dependence
- Cramping and diarrhea
- Weight loss (increased caloric intake required)
- Osteopenia (18% reduction in bone mineral content)
- Liver disease
- Immunocompromization and overwhelming post-splenectomy infections

Regional pancreatectomy- surgical removal of the pancreas, adjacent lymph nodes, some of the blood vessels that supply the organ, and possibly the lower stomach. There is evidence to suggest that a regional pancreatectomy may lead to longer survival than a total pancreatectomy. It is especially effective for small tumors at the head of the pancreas.

Potential side effects or regional pancreatectomy:

- Lower extremity lymphedema
- Insulin dependence

- Cramping and diarrhea
- Weight loss (increased caloric intake required)
- Osteopenia (18% reduction in bone mineral content)
- Liver disease
- Immunocompromization and overwhelming post-splenectomy infections

Exercise and diabetes – exercise can help the patient/client to improve their blood sugar control while reducing their risk of heart disease and nerve damage. With diabetes mellitus there is a serious condition that can occur from the build up of fat in the arteries because of the improper utilization of carbohydrates. When this happens there can be severe complications such as atherosclerosis, high blood pressure, and myocardial infarctions. It is also responsible for the deposition of fat in the blood vessels. Exercise will help to keep the weight down and properly utilize calories, thus reducing this accumulation of fat in the arteries and blood vessels and reducing these other health risks.

Obesity is a big concern, particularly with Type II diabetes. When a person is obese, the reception of the glucose-utilizing cells to insulin decreases. Therefore, the glucose is not utilized properly and remains in the blood. When people lose weight through an exercise program diabetes can reduced and sometimes even cured.

Diabetes can impair blood flow in the feet and lead to nerve damage. Without proper care, a small injury can develop into an open sore that may be difficult to treat. Sometimes an amputation is necessary if the infection severely damages the tissue and bone. Because of ***peripheral neuropathy*** (the network of nerves are damaged and sensation is reduced. Balance and coordination can be negatively impacted and there may be muscle weakness and loss of reflexes, especially in the ankles, leading to changes in the way a person walks) you can develop a blister or sore on the feet without even realizing it. It also becomes an area of concern when having a pedicure and/or shaving, if the skin is nicked or cut and blood is drawn, because of the possibility of bacteria getting into the wound.

Blood sugar needs to be tracked before, during, and after exercise. The records will reveal how their body is responding to the exercise and help to prevent dangerous blood sugar fluctuations. When the patient/client is taking Insulin or any other medication that can cause low blood sugar, they should check their blood sugar thirty minutes prior to exercising. This will help to determine if the blood sugar level is stable, rising, or falling, and if it will be safe for them to exercise. Here are some general guidelines to follow based on their blood sugar levels:

Lower than 100 – blood sugar may be too low to exercise safely. They should eat a small snack of simple carbohydrates such as fruit, juice, or crackers prior to beginning a workout.

100-250 – this is a safe blood sugar range with virtually no restrictions.

Above 250 – this is a time when it would be advisable to check the urine for keytones (substances made when your body breaks down fat for energy). If there are excess keytones, it means that the body doesn't have enough Insulin to control the blood sugar levels. If the patient/client exercises when there is a high level of keytones, they risk ***ketoacidosis*** - a serious complication of diabetes that requires immediate treatment. It is best for them to wait to exercise until they have a low level of keytones in their urine.

Some important points to remember:

- Exercise should be done in moderation with a proper warm-up and cool-down to help the body adapt to the physical strain.
- Exercise is best when done at the same approximate time every day. This will help to control blood sugar.
- Shoes should be well-fitting and designed for the appropriate activity they are participating in. Always wear socks and be alert for blisters
- If they already have foot problems such as peripheral neuropathy, consider low impact activities.

Biliary bypass surgery – this procedure is performed when an inoperable tumor is blocking the duodenum, causing bile fluid to build up in the gallbladder, or when it is blocking the common bile duct that passes through the pancreas. In the first situation, the surgeon bypasses the bile duct to the jejunum, the section of the small intestine below the duodenum. In the next situation, a catheter tube is surgically implanted in the bile duct. The fluid then drains into a small external bag. A nerve block is performed during the operation to relieve pain.

Chemoradiation (palliative treatment) – both chemotherapy and radiation can ease the pain of unresectable pancreatic cancer. Used together, they may also prolong patients' lives. The standard drug that is used is gemcitabine (Gemzar). It doesn't seem to shrink cancer cells, but it has been shown to prolong life in patients who have incurable pancreatic cancer. A new technique for patients who have extensive local spreading of their pancreatic cancer is intraoperative radiation therapy. The surgeon removes the tumor and while still in the operating room, the patient receives a high dose of radiation therapy. Because the radiation therapy is administrated during the surgical procedure, and can be delivered to a precisely defined area around the pancreas, it is possible to use a higher-than-usual dose of radiation while sparing nearby healthy organs (which are temporarily pushed aside during the procedure).

Potential side effects of chemoradiation:

- Fatigue
- Nausea, vomiting
- Diarrhea
- Abdominal pain
- Weight gain/loss
- Appetite loss
- Bowel obstruction if surgery is performed
- Digestive problems, malnutrition
- Peripheral neuropathy of the feet and hands

Gastric Bypass – if the tumor is blocking the flow of food from the stomach, the stomach may be sewn directly to the small intestine so the patient can continue to eat normally.

Potential side effects of gastric bypass:

- Dumping syndrome - with most or all of the stomach missing, the food spills into the intestine too rapidly. In *late dumping syndrome* the small intestine is forced to absorb larger amounts of food than normal, driving up the concentration of sugar in the circulation. The pancreas produce excess insulin to regulate the blood glucose level. Patients may feel weak or tired several hours after eating from a drop in blood pressure. They may also have a headache, sweating, anxiety, and/or tremors. *Early dumping syndrome* can take place several minutes after eating. Blood pressure increases, but blood flow to the intestine decreases, Symptoms include an irregular or rapid heartbeat, dizziness, shortness of breath, flushed skin, vomiting, abdominal cramps, and diarrhea. The smaller the remaining stomach, the worse the symptoms. The symptoms usually subside within 3-12 months, but in some patients the condition may become chronic. Patients can control their symptoms by eating frequent, smaller meals, low in carbohydrates. Fluids should be consumed between meals rather than accompanying them.

- Long-term deficiencies of vitamin B12, folate, and iron
- Possible weight gain following surgery

Biliary stent placement – a small plastic or wire-mesh tube is used to hold the tumor open so that bile is able to travel through. Two methods are used; endoscopic approach and percutaneous stent placement. In the endoscopic approach a flexible viewing scope is fed down the throat until it reaches the common bile duct. The stent is fed through the instrument and placed in position. This procedure is done with a topical anesthetic sprayed at the back of the throat, and intravenous sedation. In the latter, patients are put to sleep and their right lower side is numbed with local anesthetic.

Celiac block – this is an effective way to control pain by injecting alcohol into the tangle of nerves that serve the pancreas, thereby destroying them. The procedure, also called a chemical splanchnicectomy, is often performed at the same time as surgery to bypass an obstructive tumor in the abdomen.

Immunotherapy / tumor vaccines – pancreatic cancer is one of the most deadly cancers. Trials are being conducted around the country testing various immunotherapies designed to incite patient's immune systems to attack the tumor. They may be administered alone or in combination with chemotherapy. Most of these trial treatments are still used only for patients considered beyond cure.

Cancer of the Cervix

Today we are fortunate to have the *Pap smear* as a screening test not only for malignant cervical cells, but for those that may progress to cervical cancer as well. A suspicious *pap smear* is followed by a cervical biopsy to come up with a definitive diagnosis of cancer. Sometimes, however, the results of a biopsy may be inconclusive and necessitate removal of a larger tissue sample. Determining the extent of the cancerous growth is somewhat primitive compared to staging for other types of cancers. A pelvic exam will determine how big the tumor is and if it has invaded any adjacent tissues. A CT scan will more than likely be conducted to determine metastasis in the lymphatic system. Some gynecologic oncologists believe in staging the disease by sampling pelvic and abdominal lymph nodes which helps to determine the exact area to receive radiation therapy, while others don't think that a lymph node dissection should be a "routine" procedure for staging patients. There are usually no visible signs or symptoms of cervical cancer. As the cancer grows there may be unusual bleeding or discharge. You may have longer menstrual periods, a heavier flow, bleeding between periods and after intercourse, or bleeding after menopause. The bleeding is usually bright red and unpredictable as to when it appears, its amount, or its duration. Although these symptoms may not be cancer, they should be checked by the doctor.

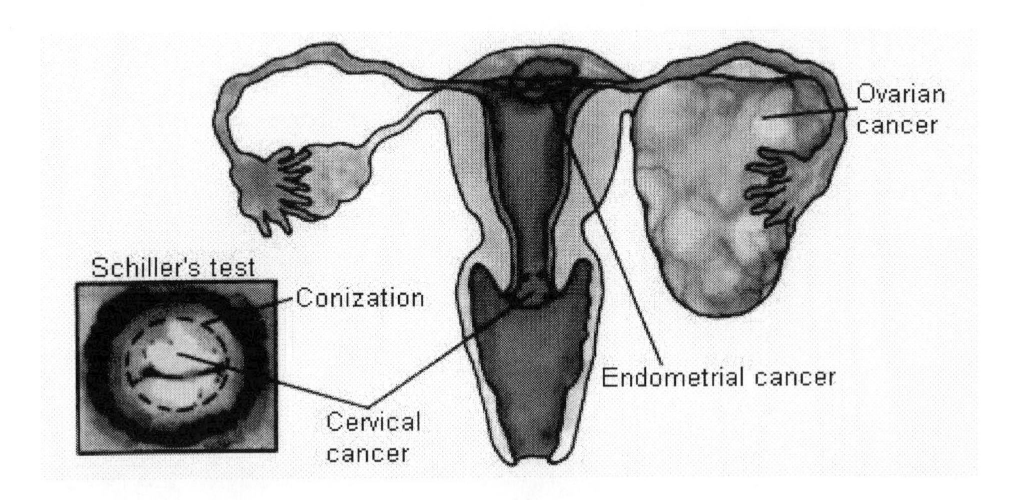

Treatment options:

Colposcopy - if the pap smear indicates that there are suspicious cells. The doctor will usually examine the vagina and cervix with a colposcope. It is essentially a microscope on a stand that gives a lighted, magnified view showing greater detail than can be seen with the naked eye. At the same time the doctor may also remove tissue for a biopsy.

Loop electrosurgical excision procedure (LEEP) – an office procedure that uses an electrically charged wire loop to slice off the outermost layer of the cervix.

Potential side effects of LEEP:

- Abdominal cramping
- Bleeding

Cryosurgery – applies extreme cold to cancer cells in order to destroy them.

Potential side effects of cryosurgery:

- Cryosurgery for cervical intraepithelial neoplasia has not been shown to affect a woman's fertility, but it can cause cramping, pain, or bleeding.
- When used to treat skin cancer (including Kaposi's sarcoma):
    - Scarring and swelling can occur
    - If nerves are damaged, loss of sensation may occur
    - Rarely, it may cause a loss of pigmentation and loss of hair in the treated area.
- When used to treat tumors of the bone:
    - It may lead to the destruction of nearby bone tissue and result in fractures
    - These effects may not be seen for some time after the initial treatment and can often be delayed with other treatments.

Surgical conization – surgery to remove a cone-shaped section of tissue from the cervix and cervical canal with either a scalpel (cold knife conization) or a laser. The cone biopsy often doubles as a diagnostic and therapeutic measure in that if the margins of the cone-shaped specimen are negative, no further treatment is warranted.

Potential side effects of surgical conization:

- Abdominal cramping
- Bleeding

Laparoscopic hysterectomy: Requires three or four small incisions no more than half an inch long. A viewing instrument is inserted through an incision in the navel. The surgeon then severs the uterus, tubes, and ovaries and takes them out through one of the other incisions. If necessary, a lymph node dissection can also be done this way.

Advantages over the traditional hysterectomy:

- Minimally invasive procedure
- Causes far less discomfort for patients
- Majority will go home the same day of surgery - worse case scenario is a forty-eight hour hospital stay

- One to two week recovery period

Total hysterectomy: Surgical removal of the uterus and cervix. The ovaries and fallopian tubes are not removed unless there is another medical reason to do so. Conventional abdominal hysterectomy leaves a five-inch vertical scar from just below the belly button to the pubic bone; or, if it's a bikini incision, from side to side. The recovery period is generally 6-8 weeks.

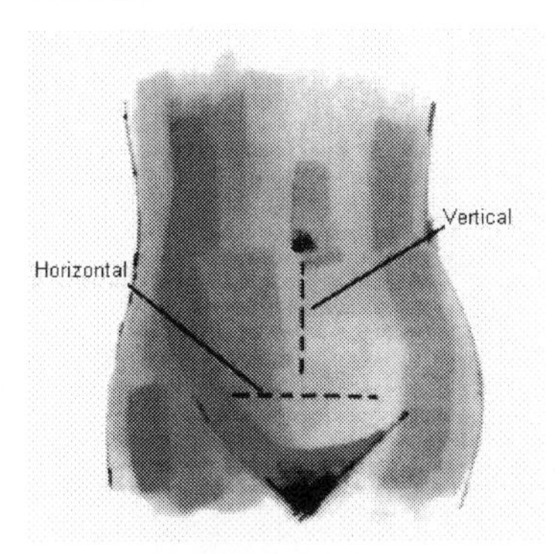

Radical hysterectomy: Surgical removal of the uterus, cervix, and part of the vagina. Area lymph nodes are removed. The ovaries and fallopian tubes are not removed unless there is another medical reason to do so.

Potential side effects for all three procedures:

- Lower extremity lymphedema (if dissection was performed)
- Instant menopause
- Infertility
- Unusual bleeding
- Wound infection
- Damage to the urinary or intestinal systems
- Pelvic pain
- Difficulty with bowel movements and urination

Radical trachelectomy – allows for some women of child bearing years to be treated without losing their ability to have children. This procedure removes the cervix and the upper part of the vagina. They place a "purse-string" stitch that acts as an artificial opening of the cervix. Nearby lymph nodded are removed laparoscopically. The operation can be done through the vagina or the abdomen.

Potential side effects:

- Lower extremity lymphedema

Pelvic exenteration – surgery to remove the cervix, uterus, ovaries, fallopian tubes, vagina, bladder, lower ureter tubes, rectum, anus, pelvic floor, and usually the pelvic lymph nodes. Because the bladder and rectum are removed, artificial openings are made to empty both urine and stools. A new vagina can also be created through plastic surgery. Recovery from this procedure takes a long time. Most women don't begin to feel like themselves for about 6 months after surgery; and for some up to a year or two.

Potential side effects of pelvic exenteration:

- Lower extremity lymphedema
- Instant menopause - menopausal symptoms
- Irritation at stoma site
- Sexuality is severely affected

Sentinel Lymph Node Biopsy - using a blue dye and a special radioactive substance that can be traced using imaging techniques, doctors can identify during surgery the first lymph node (the sentinel node) to which cancer cells would travel after leaving the uterus. This technique is called intraoperative lymphatic mapping, or sentinel node mapping. If this node is free of cancer cells, the goal is to avoid removing additional lymph nodes. If the node does contain cancer cells, then the surgeon continues to remove additional lymph nodes for further examination. Sentinel node mapping may help avoid the unnecessary removal of lymph nodes in some women, leaving these nodes in place to continue their important role in draining fluids and fighting infection.

Potential side effects:

- Lower extremity lymphedema

Hormonal replacement therapy - if a patient is relatively young and receives radiation therapy, they will have to take hormone replacement indefinitely in order to replace estrogen and prevent the crippling effects of osteoporosis.

Potential side effects of hormonal replacement therapy:

- Osteoporosis and bone fractures because they remove all estrogens from a postmenopausal woman.

- Sweating
- Fatigue
- Appetite changes
- Nausea and vomiting
- Aches and pains (headaches, bone pain, chest pain)
- Instant menopause - menopausal symptoms
- Cardiovascular problems (such as blood clots) occur at similar or lower rates than Tamoxifen

Seed-implant therapy (brachytherapy) – is considered internal radiation. A radioactive substance is applied directly to the tumor. It is used in combination with external beam radiotherapy

Radiation therapy – radiotherapy will destroy the ovaries, inducing immediate menopause. When estrogen production is halted abruptly, common menopausal symptoms such as vaginal dryness, hot flashes, night sweats, and irritability, become more severe. This is a major consideration for younger women, when deciding between a radical hysterectomy, which leaves the ovaries intact, and radiation therapy. Radiation is usually chosen for elderly patients, women who are extremely obese, or those that have other medical problems that would make surgery more risky.

Extended-field radiation – is used when an invasive cervical tumor measures 4 cm or more. The radiation field may be extended to include the *para-aortic lymph nodes* in the upper abdomen, even if there is no evidence of nodal involvement.

Potential side effects of radiation therapy and extended-field radiation therapy:

- Frequent or painful urination
- Diarrhea
- Abdominal cramping
- Vaginal itching, burning, dryness
- Bowel obstruction (from internal radiation)
- Painful sexual intercourse due to the narrowing of the vagina
- Infertility

- Menopause
- Lower extremity lymphedema

Chemoradiation – a combination of chemotherapy and radiation is used simultaneously, to replace surgery, once cervical cancer has spread to surrounding tissues. Chemotherapy is given during the four to five weeks of external-beam radiation. A week after the last treatment, patients are checked into the hospital for intercavitary implants in which a capsule containing a radioactive substance is inserted directly into the cervix and left in place for one to three days. Typically, the treatment is repeated the following week.

Potential side effects of chemoradiation:

- Fatigue
- Nausea, vomiting
- Diarrhea
- Abdominal pain
- Weight gain/loss
- Appetite loss
- Peripheral neuropathy of the feet and hands
- See previous radiation side effects

Cancer of the Uterus

Endometrial carcinoma is the third most common malignancy among women, however, it has one of the highest cure rates. Unlike cervical cancer, there is no noninvasive stage with endometrial cancer. In ***atypical uterine hyperplasia***, normal endometrial cells multiply out of control until they build up and cause the uterine wall to thicken. Depending on the degree of the hyperplasia, a woman will have anywhere from a 5-30 percent chance of developing endometrial cancer over the five to ten years. Although atypical hyperplasia is not cancer, it is treated as though it were through either surgery or hormonal treatment, depending on whether or not the patient is menopausal, or unable to bear children. Endometrial cancer usually affects menopausal women. Seventy-five percent of all cases occur after the age of 50 and only four percent before the age of 40. the most common side-effect of endometrial cancer is abnormal bleeding after menopause.

Treatment options:

Laparoscopic hysterectomy - conventional abdominal hysterectomy leaves a five-inch vertical scar from just below the belly button to the pubic bone; or, if it's a bikini incision, from side to side. Laparoscopic hysterectomy requires three or four small incisions no more than half an inch long. A viewing instrument is inserted through an incision in the navel. The surgeon then severs the uterus and takes them out through one of the other incisions. If necessary, a lymph node dissection can also be done this way. This minimally invasive procedure causes far less discomfort for patients, the majority of whom will go home the same day of surgery. Worse case scenario is a forty-eight hour hospital stay. The recovery period is considerably less; one to two weeks compared to the normal six to eight week recovery period with a standard hysterectomy.

Advantages over the traditional hysterectomy:

- Minimally invasive procedure
- Causes far less discomfort for patients
- Majority will go home the same day of surgery - worse case scenario is a forty-eight hour hospital stay
- One to two week recovery period

Total hysterectomy - surgical removal of the uterus and cervix, plus both ovaries and fallopian tubes (bilateral salpingo-oophorectomy) through an incision in the abdomen. Some lymph nodes in the pelvis and abdomen may also be removed for examination. Abdominal hysterectomy leaves a five inch scar just below the belly button to the pubic bone, or from side to side. If the uterus is removed through the vagina, it is called a vaginal hysterectomy. If a vaginal hysterectomy is done, lymph nodes can be removed laparoscopically through the abdomen. The average hospital stay is 3-7 days and there is an extensive six to eight week recovery period.

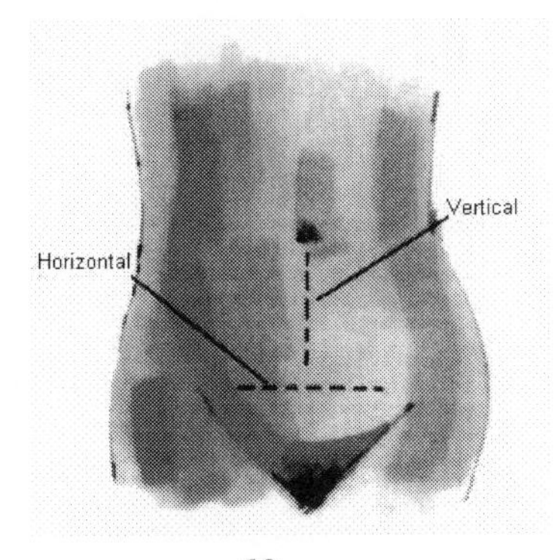

Radical hysterectomy: surgical removal of the uterus, cervix, fallopian tubes, ovaries (bilateral salpingo-oophorectomy), part of the vagina, and area lymph nodes.

Potential side effects for all three procedures:

- Lower extremity lymphedema (if dissection was performed)
- Instant menopause
- Infertility
- Unusual bleeding
- Wound infection
- Damage to the urinary or intestinal systems
- Pelvic pain
- Difficulty with bowel movements and urination

Tumor debulking – if the cancer has spread throughout the abdomen the goal is to remove as much of the tumor as possible. Tumor debulking is very helpful with other cancers, but it isn't clear yet whether it will help patients with endometrial cancer live longer.

Panniculectomy - obesity is common among patients with endometrial cancer. In certain cases the gynecologic surgeon collaborates with a plastic surgeon to combine hysterectomy and staging with a reconstructive procedure known as panniculectomy, or "tummy tuck," to remove excess skin and underlying fat in the abdominal area. This option for combination surgery is associated with better staging results and fewer complications.

Sentinel Lymph Node Biopsy - using a blue dye and a special radioactive substance that can be traced using imaging techniques, doctors can identify during surgery the first lymph node (the sentinel node) to which cancer cells would travel after leaving the uterus. This technique is called intraoperative lymphatic mapping, or sentinel node mapping. If this node is free of cancer cells, the goal is to avoid removing additional lymph nodes. If the node does contain cancer cells, then the surgeon continues to remove additional lymph nodes for further examination. Sentinel node mapping may help avoid the unnecessary removal of lymph nodes in some women, leaving these nodes in place to continue their important role in draining fluids and fighting infection.

Potential side effects:

- Lower extremity lymphedema

Chemotherapy – chemotherapy is not as commonly used to treat endometrial cancer as it is for other types of cancer. In general, chemotherapy is only used to treat endometrial cancer that ahs spread too widely to be treated effectively with surgery or radiation therapy. It is also used to treat endometrial cancer that has recurred, or come back. Hormonal therapy is usually offered as the initial treatment for metastatic tumors because its' side effects are usually mild.

Potential side effects of chemotherapy:

- Hair loss
- Dry skin or rash
- Weight gain/loss
- Fatigue
- Nausea, vomiting, or diarrhea
- Decreased appetite
- Nerve damage causing arm or leg tingling and numbness
- Muscular weakness / fatigue
- Immunocompromization
- Damage to the circulatory system and heart muscle (cardiomyopathy or accelerated atherosclerosis)
- Diabetes and osteoporosis
- Peripheral neuropathy of the feet and hands

Radiation therapy – may be used as the primary treatment when a woman is obese. The rationale for this is that obesity predisposes women to cardiovascular disease, diabetes, and other disorders that may preclude surgery. Although the survival rates are not quite as good as those of surgery, they are, in fact, encouraging. Another time that radiation would be used is when a stage III tumor has spread to the pelvic wall, at which point the cancer is declared inoperable. In stage II tumors, it may be administered before or after the total hysterectomy. Radiation may also be used to contain the tumor when it spreads or reemerges in the pelvic region.

Potential side effects of radiation therapy:

- Diarrhea/vomiting
- Low red and white blood counts
- Vaginal stenosis (scar tissue)
- Fatigue
- Sunburned look and feel in treated area
- Discomfort while urinating
- Vaginal dryness, itching, burning, and pain with intercourse
- Bowel obstruction
- Rectal bleeding
- Mucus discharge from the vagina
- Weakening of hips and pelvic bones that can lead to fracture
- Infection

Vaginal brachytherapy – a cylinder containing the radioactive pellets is inserted into the vagina; this mostly limits the radiation to the area that's in contact with the cylinder. Nearby areas such as the bladder and the rectum get minimal exposure to the radiation. The procedure is typically done 4-6 weeks after hysterectomy. There are two types, low dose brachytherapy (LDR), and high dose brachytherapy (HDR). In LDR, the radioactive pellets are usually left in place for a day at a time. The patient needs to stay immobile so the pellets don't move, so they are usually kept in the hospital overnight. Several low dose treatments may be necessary. In HDR brachytherapy, because the radiation is so intense, each dose is usually only left in place for an hour at a time. For endometrial cancer doses are usually given daily or weekly for at least three doses.

Hormonal therapy – endometrial cancer, like breast cancer, is spurred by the presence of the female sex hormone, estrogen. When the primary tumor has metastasized to or recurred in the lungs or other distant organs, patients are usually given a synthetic form of progesterone to suppress estrogen levels. These progestational agents are more likely to have an effect on endometrial tumors that contain detectable levels of progesterone and estrogen receptors.

Potential side effects of hormonal therapy:

- Osteoporosis and bone fractures because they remove all estrogens from a postmenopausal woman.
- Sweating
- Fatigue
- Appetite changes
- Nausea and vomiting
- Aches and pains (headaches, bone pain, chest pain)
- Menopausal symptoms/infertility
- Cardiovascular problems (such as blood clots) occur at similar or lower rates than Tamoxifen

Cancer of the Ovaries

Usually by the time ovarian cancer is detected, it has spread to other organs within the pelvis. Ovarian tumors are never biopsied with a needle, or cut into, because doing so could allow cancerous cells to escape. Most ovarian cancer patients are past menopause, thus making a hysterectomy a viable option when necessary. As with pancreatic cancer, ovarian cancer is considered silent and deadly. Often there are no symptoms of early ovarian cancer, which makes it difficult to detect at an early stage when there is the greatest chance for a cure. An ovarian tumor can grow for some time before pressure or pain can be felt. Symptoms, when they do occur, can include abdominal swelling or bloating, discomfort in the lower part of the abdomen, feeling full after a light meal, nausea or vomiting, lack of appetite, gas or indigestion, unexplained weight loss, diarrhea, constipation, or frequent urination, shortness of breath, and bleeding that is not part of a regular menstrual cycle.

Treatment options:

Laparotomy – To explore the abdomen for ovarian cancer, a surgery called laparotomy is necessary. In laparotomy, an incision is made in the abdomen, the area is examined, cancerous tissue is removed, and if necessary, fluid is drained from the abdominal region. It is during laparotomy that tumor debulking is performed

Tumor debulking – when a malignant ovarian mass has spread to other organs, the surgeon will remove the reproductive organs, the omentum, and the lymph nodes, and cut out, or *debulk*, as much of the tumor as possible. The goal, of course, is to eliminate any visible traces of cancer. If the tumor left behind measures half and inch or less, it's called *optimal residual cancer.* Larger tumors are referred to as *suboptimal residual cancers.* Chemotherapy cannot penetrate a large, bulky ovarian tumor because the flow of the treatment is blocked. By removing as much of the tumor as possible through debulking, the chemotherapeutic treatment is able to penetrate the tumor more effectively. This means that the tumor will be much more responsive to chemotherapeutic treatment, which improves treatment success and potentially adds years to a patient's survival.

Potential side effects for tumor debulking:

- Lower extremity lymphedema
- Instant menopause
- Infertility

Second-look laparotomy – exploratory surgery to look for evidence of any residual cancer following the completion of chemotherapy for stage III ovarian cancer. The abdomen and pelvis are carefully inspected, tissue samples are taken, and lymph nodes are biopsied. This operation is usually done through a vertical incision in the abdomen. It is generally recommended that before patients are scheduled for this procedure, they undergo a CT scan, a CA-125 tumor-marker blood test, and a physical exam to reveal any evidence that the cancer has withstood the chemotherapy.

Potential side effects of second-look laparotomy:

- Lower extremity lymphedema

Total abdominal hysterectomy and bilateral salpingooophorectomy, with omentectomy and lymph-node dissection – surgical removal of the uterus, both ovaries and fallopian tubes, and the omentum (an apron of fatty tissue that hangs from the stomach to the colon. Lymph nodes in the pelvis and abdomen are removed and examined for cancer.

Potential side effects of total abdominal hysterectomy:

- Lower extremity lymphedema
- Menopausal symptoms
- Infertility

Chemotherapy – is a systemic therapy that is administered orally or injected into a vein and delivers chemotherapy drugs throughout the body. Regional therapy is the administration of chemotherapy drugs directly into the region of the body where the tumor(s) is/are located. To treat ovarian cancer, a kind of regional chemotherapy called intraperitoneal (IP) chemotherapy is used. Chemotherapy drugs are injected into the internal lining of the abdominal area (called the peritoneal cavity) through catheter. This allows a high concentration for a prolonged period of time to reach the cancerous tissue, a treatment which has been shown to increase effectiveness in several randomized clinical trials. The catheter can be placed during surgery, but if done at a later time, it can be implanted laparoscopically. The catheter is usually connected to a "port" which is placed under the skin against a bony structure of the abdominal wall; typically a rib or pelvic bone.

Potential side effects of chemotherapy:

- Joint or muscle pain or stiffness
- Hand and foot rashes
- Mouth sores
- Immunocompromization
- Fatigue
- Infertility
- Increased bleeding or bruising after minor cuts
- Diarrhea
- Nausea and/or vomiting
- Numbness or tingling in the hands/fingers, or feet/toes
- Poor liver function
- Weight gain/loss

- Hair loss
- Pain at injection site or in vein
- Kidney damage
- Peripheral neuropathy of the feet and hands

Intra-abdominal chemotherapy – has recently been used to treat women with advanced ovarian cancer by combining standard intravenous chemotherapy along with chemotherapy that is injected directly into the abdominal cavity through a catheter placed at the time of the initial operation. The infusion exposes hard-to-reach cancer cells to higher levels of chemotherapy than can be reached intravenously. The intensity of the side-effects cause many women to be unable to complete a full treatment, however, this may still help women to live longer.

Potential side effects of Intra-abdominal chemotherapy:

- Abdominal pain
- Nausea and vomiting

Radiation therapy – post-operative radiotherapy is delivered in one of two ways: externally to the abdomen and pelvic region, or internally. It is rarely used as a primary treatment for ovarian cancer, but is sometimes considered after the removal of a recurrent tumor or in the treatment of a recurrence. It is not considered the most-effective treatment for ovarian cancer.

Potential side effects of radiation therapy:

- Diarrhea
- Cramping, abdominal pain
- Bowel obstruction
- Possible infertility

Cancer of the Stomach

The incidence of stomach cancer has gone down seventy-five percent since the 1930's, when the disease was the leading cause of death among men. Thanks to home refrigeration and more and more families able to afford an icebox, foods no longer have to be preserved through curing, pickling, and smoking. Additionally, fruits and vegetables, which lower the risk of stomach cancer, could be eaten all year-round. Because stomach cancer is fairly uncommon in the U.S. nowadays, doctors don't routinely screen for it anymore. Therefore, when stomach cancer is detected, it is usually in the advanced stages. Unfortunately because the symptoms can be easy to ignore, cancer of the stomach can be present for a long time before it is detected. Symptoms may include indigestion, a sense of discomfort or vague pain, fullness, bloating, or belching, slight nausea, heartburn, indigestion, or loss of appetite. If these symptoms are persistent for a period of two weeks or more, you should consult your doctor. Later signs of the disease may include a dark stool which may signal blood in the stools, vomiting, rapid weight loss, and severe abdominal pain.

Treatment options:

Endoscopic mucosal resection – the cancer is removed through an endoscope that's passed down the throat to the stomach. It is only done for cancers at a very early stage in which the risk spread to the lymph nodes is very unlikely.

Total gastrectomy – this procedure is done if the cancer has spread throughout the stomach, or if it is in the upper part of the stomach near the esophagus. It is the surgical removal of all of the stomach and, occasionally, the spleen, parts of the esophagus, the pancreas, and intestines may also be removed. In selected patients, minimally invasive surgery can be used to remove stomach tumors. The end of the esophagus is attached to part of the small intestine.

The minimally invasive surgical approach for stomach cancer may spare some patients from having to undergo unnecessary and noncurative operations, and in those patients undergoing laparoscopic surgery, it may speed up their recovery process. During these procedures, a thin, lighted tube with a video camera at its tip (called a laparoscope) is inserted through a tiny incision in the abdominal wall, and the image is projected onto a large viewing screen. Guided by this highly magnified image, the surgeon can operate through tiny surgical "ports" using specially designed surgical instruments.

Potential side effects of total gastrectomy:

- Increased heartburn and regurgitation

- Dumping syndrome - with most or all of the stomach missing, the food spills into the intestine too rapidly. In *late dumping syndrome* the small intestine is forced to absorb larger amounts of food than normal, driving up the concentration of sugar in the circulation. The pancreas produce excess insulin to regulate the blood glucose level. Patients may feel weak or tired several hours after eating from a drop in blood pressure. They may also have a headache, sweating, anxiety, and/or tremors. *Early dumping syndrome* can take place several minutes after eating. Blood pressure increases, but blood flow to the intestine decreases, Symptoms include an irregular or rapid heartbeat, dizziness, shortness of breath, flushed skin, vomiting, abdominal cramps, and diarrhea. The smaller the remaining stomach, the worse the symptoms. The symptoms usually subside within 3-12 months, but in some patients the condition may become chronic. Patients can control their symptoms by eating frequent, smaller meals, low in carbohydrates. Fluids should be consumed between meals rather than accompanying them.

- Risk of overwhelming infection if spleen is removed

Subtotal gastrectomy – is usually recommended is the cancer is only in the lower part of the stomach; occasionally being used for cancers in the upper stomach. Only part of the stomach is removed; sometimes along with the duodenum (the first part of the small intestine). Nearby lymph nodes are also removed.

Distal subtotal gastrectomy - takes out the lower portion of the stomach, then joins the remainder to the small bowel.

Proximal subtotal gastrectomy – the upper stomach as well as the lower esophagus is removed then the gullet is sewed to the stomach.

Potential side effects for all three procedures:

- Increased heartburn, abdominal pain, and regurgitation
- Dumping syndrome (see above description)
- Lower extremity lymphedema
- Vitamin deficiencies
- Blood clots
- Bleeding from the surgery
- Damage to nearby organs

- **All of the above procedures involve resecting parts of nearby tissues and organs and dissecting the regional lymph nodes.**

Gastroduodenostomy – connecting what remains of the stomach to the duodenum.

Gastrojejunostomy – connecting what remains of the stomach to the jejunum.

Chemotherapy – surgery is the only curative treatment for stomach cancer, however, chemotherapy may be given as the primary treatment for cancer that has spread to distant organs. Currently, adjuvant therapy doesn't offer any real advantage, but it does decrease the chance of local recurrence. FAM is the first multidrug regimen to demonstrate some benefit in recurrent gastric cancer and is very well tolerated. It consists of 5-FU, Adriamycin, and Mutamycin. Although used very rarely, another method used to treat stomach cancer is intraperitoneal (IP) chemotherapy. In IP therapy, chemotherapy drugs are placed directly into the internal lining of the abdominal area (called the peritoneal cavity) and are released through a surgically implanted catheter. This allows a high concentration of chemotherapy agents to reach the cancerous tissue, thereby increasing the effectiveness of treatment.

Potential side effects of chemotherapy:

- Nausea and/or vomiting
- Diarrhea, possibly with cramping
- Bowel obstruction
- Feeling of fullness after eating a small amount
- Low red and white blood cell counts
- Appetite loss
- Heart damage
- Weight gain/loss
- Peripheral neuropathy of the feet and hands

External beam radiation – uses high-energy rays to kill cancer cells in a specific part of the body, or cancer cells that may be left behind after surgery. On average, the patient usually receives treatment 5 times per week over a period of weeks or months. Radiation combined with chemotherapy, especially 5 FU, may delay or prevent recurrence after surgery. Radiation therapy can also be used to relieve the symptoms of advanced stomach cancer such as pain, bleeding, and difficulty eating normally.

Potential side effects of external beam radiation:

- Nausea and/or vomiting
- Indigestion
- Mild skin problems at the radiations site
- Diarrhea
- Fatigue
- Low red and white blood cell counts

Intensity-modulated radiation therapy - IMRT is a type of 3-D radiation therapy that targets tumors with greater precision than conventional radiation therapy. Using highly sophisticated computer software and 3-D images from CT scans, the radiation oncologist can develop an individualized treatment plan that delivers high doses of radiation to cancerous tissue while sparing surrounding organs and reducing the risk of injury to healthy tissues.

Respiratory gating - is another type of radiation therapy used to treat stomach cancer with minimal damage to healthy tissue. Because tumors and organs in the abdomen shift during breathing, precise delivery of radiation therapy to cancerous tissue can be difficult. Respiratory gating entails the delivery of radiation treatment only at certain points during a patient's breathing cycle, when the "mobile" tumors and/or regions of the abdomen are in a specific position. This approach decreases the radiation dose to the surrounding healthy tissues.

Cancer of the Liver

The liver is popular territory for secondary tumors from other organs, surpassed only by the lungs. Primary liver cancer is relatively uncommon. Often times the disease has been preceded by years of chronic liver disease; cirrhosis and/or viral hepatitis. liver tumors are classified in one of four ways:

1. Localized and resectable (operable) tumors are found in one place and can be removed.

2. Localized and unresectable (inoperable) tumors are found in one area but cannot be totally removed safely.
3. In advanced cases, cancer has spread throughout the liver and/or to other parts of the body.
4. In recurrent cases, the cancer has returned to the liver or another part of the body after initial treatment.

Treatment options:

Partial hepatectomy: Removal of the part of the liver where cancer is found. The part removed may be a wedge of tissue, an entire lobe, or a larger portion of the liver, along with some of the healthy tissue around it. The remaining liver tissue takes over the functions of the liver. This surgery is only done if all of the tumor can be removed while leaving enough healthy liver behind to function. Unfortunately, liver cancer is seldom contained, and has spread to multiple sites within the liver or other organs.

Four out of five people with liver cancer have cirrhosis. For a patient with severe cirrhosis, removing even a small amount of liver tissue around the tumor might not leave enough healthy liver behind to perform essential functions. People with cirrhosis are eligible for surgery only if the cancer is small and they still have a reasonable amount of liver function.

Potential side effects of partial hepatectomy:

- Bleeding problems
- Infection
- Pneumonia

Total hepatectomy and liver transplant: is removal of the entire liver and replacement with a healthy donated liver. A liver transplant may be done when the disease is in the liver only and a donated liver can be found. If the patient has to wait for a donated liver, other treatment is given as needed. Unfortunately, not many livers are donated to people with cancer; they are reserved for people with more curable diseases.

Potential side effects of total hepatectomy and liver transplant:

- Bleeding problems
- Infection
- Immediately after surgery: poor function of the grafted liver

After transplantation patients will be taking 7-10 different types of medications. Dosages and numbers are reduced over time and it is common by six months to be down to 1-2 medications. Patients will be taking immunosuppressants for the rest of their lives in virtually all cases. Alcoholic beverages must be avoided due to the toxicity of the liver as well as the fact that it can interfere with the metabolization of certain medications.

- Patients commonly experience side effects from the drugs used to treat or prevent rejection:
  - Cortisone-like drugs produce some fluid retention and puffiness of the face, risk of worsening diabetes and osteoporosis
  - Cyclosporine (an immunosuppressant) produces some tendency to develop high blood pressure and the growth of body hair.
  - FK-506 (an immunosuppressant) includes headaches, tremors, diarrhea, increased tension, nausea, increased levels of potassium and glucose, and kidney dysfunction.

Radiofrequency ablation: The use of a special probe with tiny electrodes that kill cancer cells. Sometimes the probe is inserted directly through the skin and only local anesthesia is needed. In other cases, the probe is inserted through an incision in the abdomen. This is done in the hospital with general anesthesia.

Chemotherapy/chemoembolization – system-wide chemotherapy has shown no real advantage to date with liver cancer. Some patients due respond, however, to intra-arterial chemotherapy. Infusing chemotherapy drugs directly into the hepatic artery delivers a high concentration of cancer-killing medicine to the tumor while the normal liver tissue only receives a partial dose. Chemoembolization uses a similar principle. The agent is injected directly into the hepatic artery, causing it to close off. This robs the tumor of its primary blood supply and traps the chemo right where it needs it. Although embolization can shrink hepatic cancers and relieve pain, the benefits are generally temporary. This procedure may not be an option for many people because it requires surgery to insert a catheter into the hepatic artery; a procedure many liver patients will not tolerate well.

Potential side effects of chemotherapy/chemoembolization:

- Hair loss
- Mouth sores
- Loss of appetite
- Nausea and vomiting
- Infections

- Low red and white blood cell count
- Easy bruising or bleeding
- Fatigue
- Peripheral neuropathy
- Long term side-effects of diabetes, damage to the heart and lungs, and osteoporosis

External beam radiation – Radiation uses high-dose x-rays to shrink tumors and kill cancer cells that may have been left behind after spreading away from a tumor that was removed by surgery. These remaining cells are often referred to as *hidden cells*, because they are not detected at the time of surgery. The procedure kills cancerous cells by permanently damaging their DNA in a way that causes them to lose their ability to function and then die.

Potential side effects of external beam radiation:

- Nausea, vomiting
- Diarrhea

Cryosurgery: a treatment that uses an instrument to freeze and destroy abnormal tissue, such as carcinoma in situ (cancer that involves only the cells in which it began and that has not spread to nearby tissues). This type of treatment is also called cryotherapy. The doctor may use ultrasound to guide the instrument.

Hyperthermia - heats the tumor (*local hyperthermia*), an organ or limb (*regional hyperthermia*), or the entire body (*whole-body hyperthermia*) to between 40 and 43 degrees Celsius, making them more susceptible to radiation therapy and chemotherapy

Potential side effects of hyperthermia:

- Discomfort or even significant local pain in about half the patients treated.
- It can also cause blisters, which generally heal rapidly.
- Less commonly, it can cause burns.

Cancer of the Brain

Half of all brain tumors are benign. While a non-cancerous lesion would not be alarming in most parts of the body, the skull provides no room to accommodate any additional mass. Therefore, even benign growths can potentiate life-threatening neurological damage, depending on their location. The tumor can interfere with whatever functions that area of the brain controls; speech, movement, and cognitive abilities, just to name a few. If the tumor puts enough pressure on one of the intracranial blood vessels, blood flow to the brain could be blocked, triggering the equivalent of a stroke. The location of the tumor will determine whether or not it can be removed surgically. There are basically two types of brain cancers; those which start in the brain, and those that metastasize to the brain from cancer in some other part of the body. Those that stat in the brain are referred to as primary brain tumors.

Another serious condition can result from the tumor impeding the normal flow of cerebrospinal fluid, causing it to build up and the brain to swell. This condition is called *hydrocephalus.* It increases the intracranial pressure which will ultimately damage the fragile brain tissue.

Types of brain tumors:

Gliomas – make up about half of all brain tumors. These tumors originate in the *glial* cells that form the supportive tissue of the CNS.

Astrocytomas – are the most common form of adult brain tumor. They take hold in the star-shaped *glial* cells called *astrocytes.* Treatment will depend on the grade. Low grade or well-differentiated tumors are slow growing and rarely metastasize. Intermediate grade, or anaplastic astrocytomas, grow more rapidly and the cells exhibit malignant traits. High grade, or glioblastoma multiforme, are very rapidly growing malignant tumors that will inevitably invade surrounding tissue.

Brain-stem gliomas – only represent about 5% of all adult brain tumors. They are usually made up of astrocytomas, but may contain other types of cells as well.

Ependymomas – occur in the ependymal cells found in the lining of the brain's hollow cavities and the central canal of the spinal cord. Eighty-five percent of these tumors are benign, but the malignant form of these tumors has a greater likelihood of spreading up and down the spine via the spinal fluid.

Oligodendrogliomas – these tumors form in *oligodendrocytes*, another type of *glial* cell that transmits nerve impulses. These cells grow so slowly that they may not be detected for years.

Mixed gliomas – these contain two or more types of *glial* cells. The *oligo* cell is more promising as far as a better outcome goes because it is highly sensitive to chemotherapy. The most malignant element will determine the course of therapy.

Meningiomas – comprise about one in five brain tumors. They materialize in the meningeal membrane that covers the brain and spinal cord. Ninety-five percent of these tumors are benign. Once they are removed, they rarely recur.

Craniopharyngioma – is a benign tumor that occurs most often in children and adolescents than in adults. It develops near the optic nerve and the pituitary gland. It may bring on visual impairment and hormonal imbalances. It may also affect the *hypothalamus*, the part of the brain that regulates temperature, hunger, and thirst.

Adult pineal parenchymal tumors – tumors in this region may be difficult to access surgically because they are deep within the brain. There are several types of tumors found near the pineal gland and they can either be very aggressive or very responsive to treatment.

Treatment options:

Needle biopsy – may be used for brain tumors in hard to reach areas within the brain. The surgeon drills a small hole, called a burr hole, into the skull. A narrow, thin needle, is then inserted through the hole, and tissue is removed using the needle, which is frequently guided by CT scanning.

Image-Guided Stereotactic Surgery - frameless stereotaxy, a very precise method of operating on deep-seated brain structures, is based on the idea that all points on the brain can be described using a three-dimensional system of coordinates. Stereotaxy allows surgeons to better plan operations in advance and provides enhanced orientation and guidance during surgery. Before the procedure, technicians attach six plastic self-adhesive dots around the patient's scalp, and an MRI is performed. At the start of surgery, the exact location of these dots is registered and used to relay the position of the patient's head to the surgical navigation system. The team directs a wand-like viewing device at the patient's brain that simultaneously projects an image onto a monitor in the operating room. The image is synchronized with the patient's MRI scan taken immediately before surgery. This image gives the neurosurgeon up-to-the-moment orientation as he or she plans an approach through healthy tissue to remove the tumor. Also, the neurosurgeon can use the viewing wand to help see through the tumor to its margins (outermost edges of the tissue that is being removed) during the operation, giving assurance that the entire tumor has been removed. Frameless stereotaxy is used in conjunction with intraoperative MRI.

Functional imaging & intraoperative brain mapping - functional imaging and intraoperative brain mapping have greatly improved the safety of brain tumor surgery. Functional magnetic resonance imaging (FMRI) uses high-speed MRI to map areas of the brain associated with vision, speech, touch, movement, and other functions, the locations of which can vary from one person to the next. The map then allows a surgeon to plan surgery precisely to avoid disrupting these important areas so as to optimize the patient's quality of life.

Many brain tumor operations are performed while the patient is awake but sedated. During intraoperative brain mapping, the neurosurgeon electrically stimulates the brain and the area around the tumor using small electrodes, while asking the patient to talk, count, look at pictures, and perform other basic tasks. This process helps surgeons locate the "eloquent" regions in the brain, which govern speech, the senses, and movement, and a "map" is created of the areas where preserving brain tissue is an absolute necessity. Surgeons can then avoid these sensitive tissues while removing as much of the tumor as possible. This kind of sophisticated brain mapping allows the neurosurgeon to remove tumors that are otherwise deemed inoperable, while maximizing preservation of the patient's normal function.

Neuroendoscopy - some procedures are now performed using neuroendoscopy, in which the neurosurgeon works through a small opening in the skull using a thin tube with a powerful lens and high-resolution video camera to see into the skull and brain. Advantages of this minimally invasive neurosurgical procedure include a small incision site, an enhanced ability to perform microsurgical procedures, and potentially less trauma to healthy tissue. Following surgery, an MRI is performed to determine the extent of tumor removal and to help plan further treatment.

Craniotomy – surgical removal of a benign or malignant brain tumor. An incision is made in the scalp and the tissue is peeled back to reveal the skull. A special saw is used to cut out a plate of bone. Then the outer membrane, the *dura mater*, is cut open to expose the brain's surface. After all of most of the mass is removed, the *dura mater* is sewn back together, the piece of skull sutured back into position, and the scalp closed. There are certain times in which the bone is not replaced; the muscles in the back of the head are strong enough to adequately protect the brain.

Potential side effects of all procedures:

- Damage to normal brain tissue

- Edema
- Weakness
- Coordination problems
- Personality changes
- Difficulty in speaking and thinking
- Seizures
- Most of the side effects of surgery lessen or disappear with time
- Treatment with steroids to reduce swelling in the brain may cause increased appetite and weight gain.
- Steroids can also cause restlessness, mood swings, burning indigestion, and acne.
- Swelling of the face and feet is common.

Evoked potential electrophysiological mapping – during surgery, small electrodes are used to stimulate nerves and measure their electrical responses, or *evoked potential.* By establishing the function of specific nerves in each patient, the surgeon can identify the critical areas of the brain to avoid.

Functional image-guided surgery – this procedure is carried out prior to the operation during a special MRI scan. The patient is asked to perform certain repetitive activities. The parts of the brain that are responsible for those functions will demonstrate heightened activity, which the scan will convert to an image. The neurosurgeon will have a map that will direct him to the mass and around sensitive areas.

Ultrasonic aspiration – using a hand-held ultrasonic aspirator will produce high-frequency sound waves that cause the lesion to vibrate and break apart while leaving nerves and blood vessels intact. The instrument simultaneously "vacuums" the remaining fragments of the tumor.

Laser surgery – lasers may be used with stereotactic localization to vaporize tumor cells that are inaccessible.

Radiation therapy – most brain lesions are very responsive to external-beam radiation. For brain tumors that are too risky to operate on, it is the chosen treatment. Usually, radiation is administered after surgery to destroy any remaining tumor tissue.

Potential side effects of radiation therapy:

- Fatigue
- Headaches
- Skin irritation
- Hair loss
- Difficulty hearing due to middle-ear congestion
- Small risk of impaired intellectual functioning, particularly when radiation is combined with chemo.

Stereotactic radiosurgery - by surgically implanting radioactive pellets into the tumor site, a lethal dose of radiation is delivered, minimally affecting nearby brain tissue. This procedure is done by creating a small hole in the skull in which the *dionuclides* (pellets) are placed and subsequently removed after 6 or 7 days. Using three-dimensional guidance, the machine sends multiple beams of high-dose radiation to the tumor, blasting it from different angles. Next, the patient is repositioned on the table with a five-hundred-pound helmet containing 201 holes – one for each beam of radiation delivered by the *gamma knife.* This procedure seems to be most effective for small primary and metastatic brain tumors that measure less than 4 cm in diameter, and for those tumors that are considered inoperable.

Potential side effect of stereotactic radiosurgery:

- Nausea
- Neck stiffness
- Pain at the pin sites
- Radiation injury to brain tissue surrounding the target that may cause swelling 3-12 months after the procedure. In most cases it is temporary and resolves itself, but some patients may need steroid medications to control persistent swelling.

Chemotherapy – unfortunately has demonstrated only minimal effectiveness against brain cancer. It is almost impossible to get sufficient amounts of the drug to the cancer due to the *brain-blood barrier*, a highly efficient security system of vessels and cells that filters out substances in the circulation, including chemotherapeutic drugs.

Interstitial chemotherapy – dissolvable wafers containing chemotherapy drugs are implanted directly into the tumor bed. Over the course of several weeks, it slowly delivers one thousand times the normal dose of anti-cancer drugs to the affected area of the brain, but without the awful symptoms generally associated with systemic chemotherapy.

Implantable infusion pump – a steady stream of chemotherapy drugs are administered for days or weeks through a pump that is surgically implanted in the skull.

Intra-arterial chemotherapy – chemotherapeutic drugs are injected directly into an artery leading to the brain, enabling more of the drug to reach the tumor.

Intrathecal chemotherapy – chemotherapeutic drugs are injected into the cerebrospinal fluid. The needle is either inserted in the central canal in the lower back, or the patient has a small repository called an *ommaya reservoir* surgically implanted beneath the scalp. The chemo is channeled through a catheter tube to one of the ventricles in the cerebrum. The major drawback is that most of the chemotherapeutic agents are too lethal to be delivered into the cerebrospinal fluid and the fluid is not able to distribute the drugs any deeper than one to two millimeters into the brain.

Blood-brain barrier disruption – a sugar called *mannitol* is injected through a catheter that is inserted through the groin and fed up a vessel that supplies the brain. For up to 30 minutes, the solution causes the capillaries to shrink and separate so that the anti-cancer drugs can slip through. After thirty minutes, the cells expand again and the blood-brain barrier seals shut once again. The procedure allows the brain to receive 100 times the normal dose of chemotherapy with far less risk than high-dose systemic chemotherapy which also wipes out the bone marrow

Steroid therapy / shunt surgery – a short course of steroid medication may be prescribed for swelling of the brain tissue, or *edema.* When the mass obstructs one of the ventricles, causing it to swell, it is known as *hydrocephalus*. Occasionally a *shunt* must be implanted in a ventricle to relieve the *increased intracranial pressure (IICP)* within the skull. The excess fluid is drained through a tube that runs beneath the scalp and down through the abdominal cavity, or one of the chambers of the heart. The shunt may be removed after the craniotomy, or left in place permanently.

Seizure medications – the location of a brain tumor will determine a patient's predisposition to convulsions. *Anticonvulsant medications* can enable brain-tumor patients to go through life free of seizures.

Cancer of the Testicle

Cancer of the testicle is one of the leading malignancies among men aged fifteen to thirty-four, however, it is not at all restricted to that age group. It is one of the most curable forms of cancer if treated appropriately. Even when the tumor has metastasized, 75% of patients can be cured. Symptoms of testicular cancer may be a lump in the testicle, painless enlargement of the testicle, dull ache in lower abdomen or groin, heaviness in scrotum, sudden collection of fluid in scrotum, dragging feeling in scrotum, and tenderness or enlargement of breasts.

Treatment Options:

Radical inguinal orchiectomy – surgical removal of a cancerous testicle through an incision in the groin. It is both diagnostic and therapeutic.

Potential side effects of radical inguinal orchiectomy:

- Retrograde ejaculation (semen is discharged into the bladder instead of out of the penis)

Retroperitoneal lymph node dissection – surgical removal of the retroperitoneal lymph nodes. Because many nerves necessary for ejaculation and erection are removed, it may alter sexual ability and function.

Potential Side effects of retroperitoneal lymph node dissection:

- Lower extremity lymphedema
- Impotence
- Infertility
- Retrograde ejaculation (semen is discharged into the bladder instead of out of the penis)

Chemotherapy – uses anti-cancer agents to systemically destroy cancerous cells. It may also be used to shrink tumors, either to reduce pain, or make surgery easier. Chemotherapy is a systemic treatment that travels through the bloodstream and kills cancerous cells anywhere in the body by interfering with cell growth and division.

Potential side effects of chemotherapy:

- Hair loss
- Mouth sores

- Dry skin or rash
- Weight gain/loss
- Immunocompromization
- Easy bleeding or bruising
- Fatigue
- Nausea, vomiting, or diarrhea
- Decreased appetite
- Muscular weakness / fatigue
- Damage to circulatory system and heart muscle (cardiomyopathy or accelerated atherosclerosis)
- Diabetes and osteoporosis
- Sterility for one and a half to two years after treatment
- Peripheral neuropathy of the feet and hands
- Hearing loss, sensitivity to cold or heat, and kidney damage with Cisplatin
- Damage to the lungs, shortness of breath and trouble with physical activity with Bleomycin

External beam radiation – radiation uses high-dose x-rays to shrink tumors and kill cancer cells that may have been left behind after spreading away from a tumor removed by surgery. These remaining cells are often referred to as hidden cells, because they are not detected at the time of surgery. The procedure kills cancerous cells by permanently damaging their DNA in a way that causes them to lose their ability to function and then die.

Potential side effects of external beam radiation:

- Nausea and vomiting
- Diarrhea
- Reduced fertility

Autologous bone-marrow or peripheral-blood stem-cell transplantation – once the cancer is in remission, a solid tumor may benefit from autologous bone-marrow or peripheral-blood stem-cell transplantation as a complement to high dose chemotherapy. There's very little chance of rejection since autologous BMT gives you back your own stem cells and Graft Versus Host Disease (the donor marrow perceives the host as foreign and attacks it) reactions are extremely rare. Because there is no need for immunosuppressant drugs, the immune system rebuilds much more rapidly.

Potential side effects of autologous BMT or peripheral-blood stem-cell transplantation:

- Graft rejection
- Graft failure
- Infection
- Immunocompromization for at least one year

Cancer of the Small Intestine

There are four types of small-intestine cancer; leiomyosarcoma, lymphoma, adenocarcinomas, and carcinoid tumors. Carcinoid tumors only make up for roughly 20% of all small-intestine tumors, but are the most curable type. Two in five carcinoid tumors grow in the small bowel. One in five of these will spread, usually to the liver. Metastatic carcinoids that release large quantities of hormones and other substances may be accompanied by an abundance of symptoms referred to as carcinoid syndrome: flushing of the face, diarrhea, abrupt drops in blood pressure, abdominal edema, and bronchospasms. Most people who have Carcinoid syndrome also have liver metastasis. Interestingly, even in patients with metastatic disease, the cancer grows very slowly. Watchful waiting may be recommended instead of immediate treatment.

Treatment options:

Local Excision – surgical removal of the cancer through an endoscope inserted down the throat.

Small bowel resection – surgical removal of the cancerous part of the small intestine. A *duodenectomy* resects part or all of the duodenum (the upper portion), a *jejunectomy*, part or all of the jejunum (the middle section), and *ileectomy*, part or all of the ilium (the portion that leads to the colon). Regional lymph nodes are usually removed as well.

Potential side effects:

- Lower extremity lymphedema
- Incisional hernia
- Narrowing of the stoma opening
- Blockage of the intestine from scar tissue

Chemotherapy – uses anti-cancer agents to systemically destroy cancerous cells. It may also be used to shrink tumors, either to reduce pain, or make surgery easier. Chemotherapy is a systemic treatment that travels through the bloodstream and kills cancerous cells anywhere in the body by interfering with cell growth and division. In this case it enters the treatment plan when the tumor cannot be completely removed surgically or when it has spread to the regional lymph nodes or to distant sites.

Potential side effects of chemotherapy:

- Hair loss
- Dry skin or rash
- Weight gain/loss
- Fatigue
- Immunocompromization
- Nausea, vomiting, or diarrhea
- Decreased appetite
- Muscular weakness / fatigue
- Decrease blood cell counts which may result in anemia, infection, and clotting problems
- Damage to circulatory system and heart muscle (cardiomyopathy or accelerated atherosclerosis)
- Diabetes and osteoporosis
- Peripheral neuropathy of the feet and hands

External beam radiation – radiation uses high-dose x-rays to shrink tumors and kill cancer cells that may have been left behind after spreading away from a tumor removed by surgery. These remaining cells are often referred to as hidden cells, because they are not detected at the time of surgery. The procedure kills cancerous cells by permanently damaging their DNA in a way that causes them to lose their ability to function and then die. Radiotherapy is used mainly for inoperable cancers of the small intestine.

Potential side effects of external beam radiation:

- Nausea and vomiting
- Diarrhea
- Abdominal pain
- Straining while urinating or defecating

Cancer of the Lip and Oral Cavity

The two most important factors in the outcome of the disease are the tumor location and depth. Cancer of the tongue and the floor of the mouth are most likely to spread to the neck. The cure rates for cancers of the lip, hard palate, and upper gum, go as high as 100%. The first sign of oral cancer is usually a bump in the mouth, a canker sore that doesn't heal in 2-4 weeks, or trouble swallowing or chewing. Many oral cancers get detected during routine dental examinations. Dentists might detect white patches that won't rub off, referred to as *leukoplakia*, or *erythroplakia*, a red spot with a velvety surface (much more of a concern than the white ones).

Treatment options:

Surgical excision – surgical removal of the cancer and some of the healthy tissue. Depending on the tumor's size and site, a more extensive operation may be required

Glossectomy – surgical removal of the tongue.

Potential side effects of glossectomy:

- Impaired speech
- Difficulty swallowing
- Difficulty chewing – often limited to liquids
- Malnourishment – may require tube feedings

Hemiglossectomy – surgical removal of part of the tongue

Potential side effects of hemiglossectomy:

- Impaired speech
- Difficulty swallowing
- Difficulty chewing – often limited to liquids
- Malnourishment - may require tube feedings

Hemimandibulectomy-surgery to remove the lower jaw on one side of the mouth.

Potential side effects hemimandibulectomy:

- Difficulty chewing

Laryngectomy – surgery to remove the voice box (larynx).

Potential side effects of laryngectomy:

- Must learn new ways to produce sound for speech

Neck dissection - surgery to remove the cervical lymph nodes in the neck. Whenever possible, only the cancerous nodes and a small margin are removed, but sometimes a radical dissection is required. This procedure is often followed by radiation.

Potential side effects of neck dissection:

- Lymphedema in the neck and face

Radical neck dissection – removes a block of tissue on one side of the neck from the collarbone up to the lower jaw and from the front of the neck around to the back. The jugular vein is also removed and the entire procedure is generally followed by radiation.

Potential side effects of radical neck dissection:

- Lymphedema in the neck and face

Reconstructive surgery – it may be necessary to reconstruct the tongue, the larynx, and the upper and lower jaw. This procedure harvests healthy bone or tissue from other parts of the body to fill the defect that was left by the tumor, or to replace the organ that was removed. Oftentimes, two years will pass between the time surgery is begun and reconstruction has been completed.

Radiation therapy – is usually able to remove 85% of stage I and II oral cancers. Radioactive implants have been found to be superior for tumors of the tongue, floor cheek, and lip. Often time's radioactive implants will be combined with external beam radiation.

Potential side effects of radiation therapy:

- Fatigue
- Sensation of a lump in the throat
- Painful swallowing, difficulty eating and drinking
- Mouth sores
- Skin irritation
- Hoarseness
- Damage to the taste bud and salivary glands
- Swelling near surgical site
- Dry mouth
- Inflammation and ulcerations to the lining of the mouth and throat
- Over-production of plaque on teeth due to limited saliva production, leads to cavities
- Change in contours of the gums, causing dentures to no longer fit properly
- Damage to the tooth socket
- If the temporomandibular joint (TMJ) is in the radiation field, may have pain or difficulty opening jaw

Laser surgery/cryosurgery – either procedure may be used for in situ lesions in the oral cavity

Mohs' micrographic surgery – is being studied as a treatment for early-stage oral cancers. It removes the tumor one layer at a time. The surgeon inspects the area through a microscope to make sure that no malignant cells remain.

Hyperthermia – is being used experimentally for recurrent oral cancers. The procedure kills cancer cells by heating the body

Potential side effects of hyperthermia:

- Discomfort or even significant local pain in about half the patients treated.
- It can also cause blisters, which generally heal rapidly.
- Less commonly, it can cause burns.

Hyperfractionated radiation therapy – is being used experimentally for advanced lesions of the lip. This procedure administers accelerated doses of external-beam radiation to the cancerous site.

Cancer of the Thyroid

Thyroid cancer is the most common endocrine tumor and one of the most curable forms of cancer. Thyroid cancer affects three times as many women as men. Although it's not gender specific, a history of radiation therapy to the head or neck during childhood has been linked to the disease. Thyroid tumors may appear between 5 and 20 years after radiation treatment. There are several classifications of thyroid cancer; papillary, follicular, medullary, and anaplastic. Papillary carcinoma is slow to grow and highly treatable, even if it has already invaded the lymph nodes. More than half of the cases of papillary carcinoma will, in fact, spread to the cervical lymph nodes. Follicular carcinoma is seen more often in older men and women. It's more aggressive then papillary carcinoma, but still considered treatable in most circumstances. Medullary carcinoma is the rarest type of thyroid cancer and it is not uncommon for it to metastasize to other organs. Because the tumor will frequently invade the cervical lymph nodes, this type of cancer usually warrants a total thyroidectomy plus extensive lymph-node dissection. One in ten cases of medullary carcinoma is hereditary. Finally, anaplastic thyroid cancer is characterized by metastasis to other organs. It usually strikes people over 60. The cells are extremely aberrant, explosive in their growth, and aggressive in their pattern of spread.

Treatment options:

Lobectomy – surgical removal of either of the thyroid's two lobes. This procedure is also called a hemithyroidectomy. Because this procedure leaves part of the thyroid behind, the patient may not need thyroid hormone for the rest of their life.

Isthmusectomy – surgical removal of the isthmus, the part that connects the two wing-shaped lobes.

Near-total thyroidectomy – surgery to remove the lobe containing the tumor, the isthmus, and most of the opposite lobe. This procedure is also known as a subtotal thyroidectomy.

Total thyroidectomy – surgery to remove the entire thyroid.

Potential side effects of total thyroidectomy:

- Damage to the adjacent parathyroid glands (regulate blood calcium levels)
- Excessive bleeding or formation of major blood clot (hematoma) in the neck
- Wound infection
- Hypocalcemia
- Damage to the vocal cords
- Hoarseness
- Impeded speaking and breathing
- Hormonal dependence
- Weight gain
- Lethargy

*** Lymph nodes may be removed during any of the above surgeries.**

Radioactive Iodine Therapy - some patients may need radioactive iodine treatment after surgery. Patients with medullary thyroid cancer or anaplastic thyroid cancer are not treated with radioactive iodine. Radioactive iodine is given in either liquid or pill form. Radioactive iodine is given to destroy any normal thyroid tissue. This allows your physicians to maintain your surveillance with blood tests to check a blood marker called thyroglobulin. As the radioactive iodine travels through the body, it is able to find and destroy any thyroid cells that were not removed by surgery or those thyroid cancer cells that have spread beyond the thyroid. Usually only one or two treatments with radioactive iodine are necessary. It has been shown to improve the survival rate of patients with papillary or follicular thyroid cancer that has spread to the neck or other body parts. The benefits are less clear for patients with small, contained cancers of the thyroid. Radioactive iodine is not used to treat anaplastic or medullary thyroid cancer because neither of these cancers "take up" iodine.

For radioactive iodine to be most effective, patients must have high levels of TSH in the blood. This will stimulate thyroid tissue and cancer cells to "take up" the radioactive iodine. Following thyroid surgery, the levels can be raised by abstaining from hormone replacement for several weeks. This will result in very low thyroid hormone levels (hypothyroidism) which will then cause the pituitary gland to release more TSH. Depending on the dose of radiation, you may have to stay in the hospital for several days after treatment to prevent others from being exposed to the radiation. This will not be the case for everyone. Some people will be allowed to go home right after treatment.

Potential side effects of radioiodine therapy:

- Temporary loss of taste
- Pain swelling and tenderness of the neck and salivary glands
- Dry mouth
- Taste changes
- May reduce tear formation - leading to dry eyes
- Temporary cessation of menstruation in older women
- Decreased fertility in men
- Headache, nausea, and vomiting at high doses
- Slightly increased risk of developing leukemia in the future (very rare)
    - If cancer cells have metastasized to the bladder:
        - Frequent urination
        - Discomfort while urinating
        - Bloody urine
    - If cancer cells have metastasized to the stomach:
        - Abdominal pain
        - Nausea and vomiting
        - Appetite loss
    - If cancer cells have metastasized to the central nervous system:
        - Swelling of the brain (cerebral edema)

- Compression of the spinal cord

Hormonal therapy - Thyroid hormone therapy replaces the hormones that the thyroid gland usually produces. Thyroid hormone is also used to decrease the pituitary gland's production of thyroid-stimulating hormone (TSH). Because TSH could potentially cause any remaining cancer cells to grow, sufficient thyroid hormone is given to reduce TSH to desired levels.

Radiation therapy – radiotherapy may be used to control localized papillary and follicular tumors if the radioiodine uptake is minimal. It may be used as treatment for inoperable anaplastic thyroid cancer as well. It is also used for metastatic lesions that are unresponsive to I-131.

Potential side effects of radiation therapy:

- Sore throat
- Difficulty swallowing

Intensity modulated radiation therapy - IMRT is a kind of radiation therapy that uses computer images to show the size and shape of the tumor. Thin and precise beams of radiation are then aimed at the tumor from many angles from outside the body. IMRT targets tumors so efficiently that it leaves healthy tissue unharmed.

Chemotherapy – currently, research is being done with experimental gene therapy and new chemotherapeutic agents, however, for patients who are not eligible, traditional chemotherapy is an option. Adriamycin, the most commonly prescribed drug, can shrink tumors in about one-fourth of the patients. The combination of Adriamycin and Platinol seems to be more powerful, particularly against anaplastic thyroid cancer. Chemotherapy is used for cancers that recur in other areas of the body as well. Generally speaking, however, chemotherapy is not effective for most types of thyroid cancer.

Potential side effects of chemotherapy:

- Hair loss
- Mouth sores
- Loss of appetite
- Nausea and vomiting
- Immunocomprimization

- Easy bruising or bleeding
- Fatigue
- Peripheral neuropathy of the feet and hands
- Long term – diabetes, osteoporosis, and damage to the heart and lungs

Bisphosphonate Therapy - If thyroid cancer has spread to your bones, you may be treated with a class of drugs called bisphosphonates, which are used to prevent bone loss, reduce the risk of fractures, and decrease pain.

Cancer of the Esophagus

Although esophageal cancer is rare, the most common kinds of the disease are squamous cell and adenocarcinoma. Squamous cell usually begins in the cells that form the top layer of the esophageal lining. Adenocarcinoma most often develops in people who have Barrett's esophagus, a condition in which the lining of the esophagus near the opening of the stomach changes in response to constant stomach acids (reflux). Diagnosing esophageal cancer begins with a barium swallow, followed by an esophogoscopy. Endoscopic ultrasound may be used to actually stage the cancerous growth and look at its' depth of penetration. In order to locate distant metastasis, a CT scan may be used. It is not uncommon for these tests to give a false negative regarding lymph-node involvement. Surgery may be the only true indicator of the cancers' involvement. To minimize ambiguities, a thoracosopy may be used to assess lymph nodes ad take biopsies. Malnourishment is a frequent complication and can increase the risk of not surviving the operation. Symptoms of esophageal cancer include difficulty swallowing, sensations of pressure and burning or pain in the upper middle part of the chest, hoarseness, cough, fever, or choking. The sensations may come and go.

Treatment options:

Esophagectomy – surgical removal of all or part of the esophagus. Often times a small amount of the stomach is removed as well as nearby lymph nodes. The upper part of the esophagus is then connected to the remaining part of the stomach. Part of the stomach is pulled up into the chest to become the new esophagus. If cancer is located in the distal part of the esophagus (near the stomach), the surgeon will remove part of the stomach, the part of the esophagus containing the cancer, and about 3-4 inches of healthy esophagus. The stomach is then connected to what is left of the esophagus either in the neck or high in the chest. If the cancer is in the upper or middle part of the esophagus, most of the esophagus will be removed in order to get a "healthy margin" above the cancer. The stomach will be brought up and connected to the esophagus in the neck. In the event that the stomach can't be used to replace the esophagus, the surgeon may use a piece of the intestine instead. Moving the intestine must be done with the utmost precision because if the vessels are damaged the blood supply will be compromised and the tissue will die.

Potential side effects of esophagectomy:

- Heart attack or blood clot in the lungs or brain during or after the operation
- Patients will need to eat smaller meals several times a day due to digestive problems
- Leak where stomach is connected to esophagus
- Difficulty swallowing
- Indigestion
- Heartburn
- Regurgitation of stomach acid
- Nausea and vomiting
- Infection
- Lymphedema in the neck and face

Laparoscopy - this procedure may be used to determine if the patient will benefit from surgery. A thin, lighted tube with a video camera at its tip is inserted through a tiny incision in the abdominal wall, and the image is projected onto a large viewing screen. Guided by this highly magnified image, the surgeon can inspect the outside of the stomach and the lower esophagus, as well as remove tissue samples for biopsy and operate through tiny "ports" using specially designed surgical instruments.

Mediastinoscopy or thoracoscopy – two or three incisions are made in the chest and an endoscope is placed inside to assess the lymph nodes and to take biopsies if necessary.

Potential side-effects:

- Lymphedema of the torso

Radiation therapy and chemoradiation – giving both together proves to be superior to either alone. The combination can also be used for advanced cancers to both open that part of the esophagus and treat distant metastasis. Locally recurrent esophageal cancer often cannot be removed successfully by surgery, therefore, chemotherapy will usually be prescribed. It is usually administered before surgery, rather than after surgery. Radiation therapy may also be used to relieve pain, enhance swallowing, or treat other symptoms of esophageal cancer. Brachytherapy may be useful in shrinking the tumor so the patient can swallow with less difficulty. This works better as a palliative treatment and not as a cure because it can't treat a large area.

Potential side effects of radiation therapy and chemoradiation:

- Skin changes
- Nausea and vomiting
- Diarrhea
- Fatigue
- Painful sores in the mouth and throat
- Dry, sore throat
- Dry cough
- Pain and difficulty with swallowing, may require dilation
- Shortness of breath
- Peripheral neuropathy of the hands and feet
- Long term- diabetes, osteoporosis, and damage to the heart and lungs

Intensity-modulated radiation therapy (IMRT) – may be used to decrease toxicity to normal tissue. IMRT is a type of 3-D radiation therapy that targets tumors with greater precision than conventional radiation therapy. Using highly sophisticated computer software and 3-D images from CT scans, the radiation oncologist can develop an individualized treatment plan that delivers high doses of radiation to cancerous tissue while sparing surrounding organs and reducing the risk of injury to healthy tissues.

Radiofrequency ablation – a balloon is passed into an area of Barrett esophagus. It is inflated until the surface of the balloon is in contact with the inner lining. The high-powered energy is then used to kill the cells that are in the lining. Normal cells will then begin to replace the Barrett cells. The patient will need to be on drugs to block acid production following this procedure.

Respiratory gating - is another type of radiation therapy used to treat esophageal cancer with minimal damage to healthy tissue. Because tumors and other organs in the abdomen shift during breathing, precise delivery of radiation therapy to cancerous tissue can be difficult. RG entails the delivery of radiation treatment only at certain points during a patient's breathing cycle, when the "mobile" tumors and/or regions of the abdomen are in a specific position. This approach decreases the radiation dose to the surrounding healthy tissues. RG is particularly important for the tumors at the gastroesophageal junction, where there is a lot of motion due to breathing.

Photodynamic therapy / laser therapy – is used as a palliative treatment. A photosensitizing drug (Photofrin) is used in combination with laser light to shrink tumors that are blocking the esophagus. This treatment is also being used experimentally for treating carcinoma in situ from the inner lining.

Potential side effects of PDT and laser therapy:

- Photodynamic therapy makes the skin and eyes sensitive to light for 6 weeks or more after treatment.
    - Patients are advised to avoid direct sunlight and bright indoor light for at least 6 weeks.
    - If patients must go outdoors, they need to wear protective clothing, including sunglasses.
- Other temporary side effects of PDT are related to the treatment of specific areas and can include coughing, trouble swallowing, abdominal pain, and painful breathing or shortness of breath.

Esophageal dilation / stent placement – is used as a palliative treatment to allow patients to eat normally. The esophagus is widened by passing a series of progressively larger rubber tubes down the throat and past the narrow part of the gullet. In cases where this is ineffective, a stent may be placed there to open the obstruction.

Potential side effects of esophageal dilation and stent placement:

- Solid food may become stuck because small tubes cannot move food along with the same efficacy as the esophagus
- Patients may be restricted to a soft or liquid diet

Gastrostomy / jejunostomy – prior to chemoradiation, a temporary feeding tube may be inserted to allow for nutritional supplementation do to studies that have shown that progressive weight loss in the weeks preceding surgery can increase the risk of not surviving the surgery. A catheter implanted in the stomach is called a gastrostomy, or G tube. A jejunostomy, or J tube, goes into the jejunum, the middle part of the small intestine.

Both of these procedures can be conducted through an endoscope. Simultaneously, an intravenous catheter may be inserted for infusing chemotherapy drugs.

Cancer of the Larynx

Fortunately, no additional treatment beyond the initial biopsy may be required for carcinoma in situ of the larynx. A laser may, however, be used to eradicate any affected areas of the organs' lining. Early stage tumors can be treated with radiation alone or in conjunction with surgical procedures to remove part of the larynx. Laryngeal cancer is different in that a person can have no secondary lesion or malignant lymph nodes yet still be classified as stage IV if the tumor has encroached upon the throat or soft tissue of the neck. Supraglottic cancer is the most likely to affect the nodes. Twenty-five to fifty percent of these patients will have node-positive disease. Patients with laryngeal cancer face a substantial risk of developing a secondary cancer of the head, neck, or esophagus, particularly if they continue to smoke and drink. Men are almost nine times more likely to develop cancer or the larynx than women, although the incidence in females is now rising, possibly due to increased smoking among women. Cancer of the larynx occurs most likely in people over the age of 55. It is more common among African-Americans than among Caucasians. Symptoms depend on the size and location of the tumor. Since most tumors begin in the vocal cords, they almost always cause hoarseness or other changes in the voice. Tumors in the area above the vocal cords may cause a lump in the throat, difficult or painful swallowing, a cough that persists, a sore throat, or an earache. Tumors that begin in the area below the vocal cords, which are rare, can cause shortness of breath or harsh, noisy breathing. Large tumors may cause swollen neck glands, pain, weight loss, bad breath, and frequent choking on food. Hoarseness that lasts more than three weeks should be checked by a doctor.

Treatment options:

Endoscopic re-excision – a second endoscopic surgical procedure to excise a wider margin of cancerous laryngeal tissue. This procedure is often performed using a laser.

Cordectomy – surgical removal of one or two of the vocal chords. This procedure may be performed endoscopically, or through an incision in the neck.

Partial / supraglottic laryngectomy – surgical removal of part of the larynx. Typically, the upper portion of the voice box is removed, but not the vocal chords. A vertical hemilaryngectomy removes either the right or left side of the larynx.

Total laryngectomy – surgical removal of the entire voice box. Additionally, a permanent opening is created in the neck (stoma) and a tube is inserted in the windpipe.

Potential side effects of total laryngectomy:

- Breathing, coughing, and sneezing will need to take place through the newly created stoma
- Routine care will need to be taken of the stoma
- An *Electrolarynx* will need to be used to speak and the voice will have a monotonic sound

Thyroidectomy – is used for patients with advanced supraglottic laryngeal cancer. This procedure removes the thyroid gland.

Potential side effects of thyroidectomy:

- Damage to the adjacent parathyroid glands
- Hypocalcemia
- Weight gain
- Damage to the vocal cords
- Hoarseness
- Impeded speaking and breathing
- Hormonal dependence

Radical neck dissection – during the surgery to remove the voice box, all of the lymph nodes in the neck are removed. Additionally, the *internal jugular vein*, the *spinal accessory nerve*, and the *sternocleidomastoid muscle* are removed. If any one of these structures can be spared, the procedure is referred to as a *modified radical neck dissection.*

Potential side effects of radical neck dissection:

- Shoulder movement may be impaired

- There may be some difficulty raising the arm over the head
- Mild pain due to inflammation of the shoulder joint
- Neck may take on a sunken appearance
- Reduced strength in neck and head movements
- Severe cervical lymphedema
- Numbness of the skin of the neck and ear

Tracheoesophageal puncture (TEP) – this is often the preferred choice for restoring speech. This is a one-day surgery which can be done as long as ten years after larynx surgery. Its' purpose is to give people who find it difficult to learn esophageal speech an opportunity to regain the use of their voices. An opening is made from the windpipe into the esophagus, and a small plastic valve is inserted to provide a source of air. The patient diverts air into the esophagus and uses it to resonate in the pharynx. A speech pathologist will help the patient regain good speech.

Chemotherapy – uses anti-cancer agents to systemically destroy cancerous cells. It may also be used to shrink tumors, either to reduce pain, or make surgery easier. Chemotherapy is a systemic treatment that travels through the bloodstream and kills cancerous cells anywhere in the body by interfering with cell growth and division.

Potential side effects of chemotherapy:

- Hair loss
- Dry skin or rash
- Weight gain/loss
- Immunocompromization
- Fatigue
- Nausea, vomiting, or diarrhea
- Decreased appetite
- Muscular weakness / fatigue
- Damage to circulatory system and heart muscle (cardiomyopathy or accelerated atherosclerosis)
- Diabetes and osteoporosis

- Sterility for one and a half to two years after treatment

External beam radiation – radiation uses high-dose x-rays to shrink tumors and kill cancer cells that may have been left behind after spreading away from a tumor removed by surgery. These remaining cells are often referred to as hidden cells, because they are not detected at the time of surgery. The procedure kills cancerous cells by permanently damaging their DNA in a way that causes them to lose their ability to function and then die.

Potential side effects of external beam radiation:

- Hoarseness
- Sensation of a lump in the throat
- Sore throat
- Dry mouth
- Reduction in saliva flow, possibly leading to tooth decay

Cancer of the Throat

Cancer of the throat, or pharynx, is divided into three types based on their location; *nasopharyngeal, oropharyngeal*, and *hypopharyngeal. Nasopharyngeal* cancer is the only head and neck malignancy not associated with tobacco and alcohol use. Seventy percent of all cases are caused by the *Epstein Barr Virus.* Tumors in the throat are much more serious than those in the larynx or oral cavity because they have a higher risk of metastasis. *Hypopharyngeal carcinoma* is not only the most common of the three types, it is also most likely to spread. There may also be a mass in the neck as a result of spreading to the lymph nodes. Symptoms of hypopharyngeal carcinoma include difficulty swallowing and pain on swallowing, a sore throat that does not go away, a lump in the neck, a change in voice or pain in your ear.

Types of throat cancer

- **Nasopharyngeal carcinoma** - the nasopharynx is located behind the nasal cavity and is inaccessible through surgery. Because of this factor, radiation therapy is the standard treatment. If surgery is used at all, it may be to remove lymph nodes that have withstood radiation treatment.

- **Oropharyngeal carcinoma** – a combination of radiation and surgery are used as primary treatment for early-stage cancer of the middle throat. The oropharynx includes the base of the tongue, the soft palate and its fleshy u-shaped protuberance (uvula), and the tonsils.

- **Hypopharyngeal carcinoma** – the hypopharynx includes the pyrifom sinus, the pharyngoesophageal junction (where the throat meets the esophagus), and the rear wall of the lower throat. This cancer is usually discovered at an advanced stage because of its' silent nature.

Treatment options:

Pharyngectomy – surgical removal of part of the throat. Lymph nodes in the neck may also be removed.

Potential side effects of pharyngectomy:

- Cervical lymphedema
- Loss of some speaking ability

Laryngopharyngectomy – surgical removal of the larynx and all or part of the throat. Neck dissection is invariably performed. When the entire pharynx is removed, a new throat is reconstructed using tissue transferred from another part of the body.

Potential side effects of laryngopharyngectomy:

- Cervical lymphedema
- Loss of some speaking ability

Partial laryngopharyngectomy – sometimes part of the voicebox can be preserved in this procedure. If the entire larynx is removed, an opening can be created through the neck and into the windpipe. Patients learn to breathe through this *tracheostomy* and must learn new techniques for speaking.

Radical neck dissection – during the surgery to remove the voice box, all of the lymph nodes in the neck are removed. Additionally, the *internal jugular vein*, the *spinal accessory nerve*, and the *sternocleidomastoid muscle* are removed. If any one of these structures can be spared, the procedure is referred to as a *modified radical neck dissection.*

Potential side effects of radical neck dissection:

- Shoulder movement may be impaired
- There may be some difficulty raising the arm over the head
- Mild pain due to inflammation of the shoulder joint
- Neck may take on a sunken appearance
- Reduced strength in neck and head movements
- Severe cervical lymphedema
- Numbness of the skin of the neck and ear

Throat reconstructive surgery – if a total or near total pharyngectomy is performed, the operation includes throat reconstruction using tissue from the small intestine or forearm. The blood supply is disconnected from the original site and then reconnected at the implant site. The transplanted tissue doesn't contract like the muscles of the throat to help food go up and down toward the esophagus.

Potential side effects of throat reconstructive surgery:

- Difficulty swallowing
- Food may get lodged in throat

Gastric pull-up – replaces the throat and the entire esophagus by taking the "J" shaped part of the stomach and attaching it to the pharynx.

Glossectomy – surgical removal of the tongue.

Potential side effects of glossectomy:

- Impaired speech

- Difficulty swallowing
- Difficulty chewing – often limited to liquids
- Malnourishment – may require tube feedings

Hemiglossectomy – surgical removal of part of the tongue.

Potential side effects of hemiglossectomy:

- Impaired speech
- Difficulty swallowing
- Difficulty chewing – often limited to liquids
- Malnourishment – may require tube feedings

Chemotherapy – uses anti-cancer agents to systemically destroy cancerous cells. It may also be used to shrink tumors, either to reduce pain, or make surgery easier. Chemotherapy is a systemic treatment that travels through the bloodstream and kills cancerous cells anywhere in the body by interfering with cell growth and division.

Potential side effects of chemotherapy:

- Hair loss
- Dry skin or rash
- Immunocompromization
- Fatigue
- Weight gain/loss
- Nausea, vomiting, or diarrhea
- Decreased appetite
- Muscular weakness / fatigue
- Damage to circulatory system and heart muscle (cardiomyopathy or accelerated atherosclerosis)
- Diabetes and osteoporosis
- Peripheral neuropathy of the feet and hands

External beam radiation – radiation uses high-dose x-rays to shrink tumors and kill cancer cells that may have been left behind after spreading away from a tumor removed by surgery. These remaining cells are often referred to as hidden cells, because they are not detected at the time of surgery. The procedure kills cancerous cells by permanently damaging their DNA in a way that causes them to lose their ability to function and then die.

Potential side effects of external beam radiation:

- Sensation of lump in throat
- Fatigue
- Painful swallowing, difficulty eating and drinking
- Mouth sores
- Hoarseness
- Skin irritation
- Altered taste, bitter taste, loss of taste
- Increased sensitivity of the tongue
- Swollen salivary glands after the first few days of treatment
- Swelling near surgical site
- Dry mouth
- Change in contour of the gums, causing dentures to no longer fit properly
- Damage to tooth socket, with subsequent tooth extraction
- If the temporomandibular joint (TMJ) is in the radiation field, there may be pain or difficult opening the jaw.

Moh's micrographic surgery – is being studied for early-stage oral cancers. It removes the tumor one layer at a time. The surgeon inspects the area with a microscope to make sure that no malignant cells remain.

Hyperfractionated radiation therapy – is being used experimentally for advanced lesions of the lip. This procedure administers accelerated doses of external-beam radiation to the cancerous site.

Cancer of the Breast

Probably no disease is more dreaded by most women than breast cancer. It is the leading form of cancer among females and carries with it fear of death as well as deformity. Fortunately, through early detection, the incidence of mortality among women with breast cancer has been on a steady decline. Most tumors are detected when they are still localized to the breast, allowing for breast conserving surgery instead of a mastectomy. Symptoms of breast cancer include a lump or thickening in the breast or under the arm, a change in the size or he shape of the breast, a change in the color or feel of the skin of the breast or the skin around the nipple; this may be dimpling, puckering, or scaliness of the skin, swelling, redness, or feeling of heat in the breast.

Types of breast cancer

- **Ductal carcinoma** – about three in four breast cancers fall under this category. They originate in milk ducts, are considered cancer in its earliest stage, and are 97% curable.
- **Lobular carcinoma** – isn't generally considered cancer, but rather that a woman has a high probability of forming invasive cancer, generally a one in four chance during the next 25 years. As its name would suggest, it originates in the lobes or lobules. Of the three types of breast cancer, lobular carcinoma is the one most likely to occur in both breasts (*bilaterally*).
- **Infiltrating ductal carcinoma** - cancer that can invade other tissue. It begins in the milk duct but grows into the surrounding normal tissue inside the breast.
- **Infiltrating lobular carcinoma -** cancer that can invade other tissue. It begins in the lobes or lobules, but grows into the surrounding normal tissue inside the breast.
- **Inflammatory breast cancer** – is very rare and materializes in the lymph vessels in the skin of the breast. It resembles an infection; the skin becomes thick, raised, and red. Instead of a single lump, it usually involved the entire breast. It is a very fast growing disease and usually more prone to metastasis.

Features of Breast Cancer

Stage: The stage of a cancer, sometimes referred to as TNM, is used to determine the type of treatment a woman should receive. It is calculated based on several factors:

- Size of the tumor (T)
- Whether or not the cancer has spread to the axillary lymph nodes, and if so, to how many (N)
- Whether or not the cancer has spread to other lymph nodes of the neck or chest area
- Whether or not the cancer has spread (metastasized) to other parts of the body (M)

Tests that may be used to determine how far a breast cancer has spread include a chest x-ray and blood tests. Depending on the clinical presentation, a physician might also order a bone scan, a CT scan, a PET scan, or MRI scan to assess the extent of the disease.

Hormone Receptor Status

Breast cancer cells that have receptors for the hormones estrogen and/or progesterone are called estrogen receptor (ER) and/or progesterone receptor (PR) positive. If these receptors are not present, the cell is said to be receptor-negative.

Tumors that are receptor-positive are more likely to respond to therapy with anti-estrogen medications, which take advantage of the cancer cell's dependence on hormones for growth. The drug tamoxifen, for example, acts by blocking the estrogen receptors of a breast cancer cell.

HER2/neu Status

HER-2/neu is a gene that, when activated, helps tumors grow by producing a specific growth-stimulating receptor. Tumors that have more than normal amounts of this protein (HER2-positive) may benefit from the drug trastuzumab (Herceptin®), which blocks the growth of tumors activated by the gene, or a similar drug called lapatinib (Tykerb®).

Grade- is an evaluation of how abnormal or disorganized the cells appear when examined under a microscope. In general, a lower grade implies a less aggressive tumor.

Lymphovascular or perineural invasion- sometimes tumor cells can invade the blood vessels, or the lymph or nerve channels within breast tissue.

Gene expression profiling- oncotype DX® is a test that is used to analyze the expression pattern of 21 genes in patients with a breast tumor that is estrogen-receptor-positive and axillary-lymph-node-negative. The pattern is translated into a recurrence "score" that attempts to predict which women will benefit most from treatment with chemotherapy in addition to hormonal therapy, and which women might be safely spared chemotherapy. This test is part of a novel and increasingly popular approach called "personalized medicine" for cancer, in which the molecular features of the specific tumor are analyzed to determine optimal treatment for each individual.

Treatment options:

Lumpectomy: only the cancerous breast tissue, with a rim of normal tissue around it, is removed. Whether or not a woman can undergo a lumpectomy is determined by the size of her tumor, the size of her breast, the number of sites of cancer within the breast, and whether the patient can undergo subsequent radiation treatments, among other factors. Patients who choose lumpectomy will likely be advised to have radiation therapy to the breast area after surgery.

Potential side effects of lumpectomy:

- Increased risk of lymphedema if radiation treatment or node dissection accompanies the procedure
- Skin tightness / adhesions

Partial/segmental mastectomy: surgical removal of the cancer, a wedge of normal tissue around it, and the lining over the chest muscle below the tumor. Usually some axillary lymph nodes are removed. In almost all cases there will be a course of radiation therapy following the surgery.

Potential side effects of partial/segmental mastectomy:

- Skin tightness / adhesions
- Muscular weakness (primarily *serratus anterior*) causing muscular instability of the shoulder girdle if axillary nodes are removed
- Increased risk of lymphedema if axillary nodes are removed or radiation is performed

Total/simple mastectomy: surgical removal of the entire breast and usually a few lymph nodes that are located in the breast tissue.

Potential side effects of total/simple mastectomy:

- Skin tightness / adhesions
- Painful and difficult movement of the arm and shoulder

- Increased risk of lymphedema

Skin-sparing mastectomy: is performed to facilitate immediate breast reconstruction. Skin-sparing mastectomy incisions are smaller than those required for a modified radical or simple mastectomy. Most of the breast tissue is removed, but most of the breast skin is saved to hold and shape the reconstructed breast. In a skin-sparing mastectomy, the incision is made around the areola. Sometimes it is necessary to make another incision extending down or to the side to remove as much breast tissue as possible. Research shows skin-sparing mastectomies do not increase the risk for breast cancer recurrence in patients with early stage breast cancer. Skin-sparing mastectomies are now also commonly used for prophylactic mastectomy followed by immediate reconstruction.

Potential side effects of skin-sparing mastectomy:

- Skin tightness / adhesions
- Painful and difficult movement of the arm and shoulder
- Increased risk of lymphedema if lymph nodes are removed
- Muscular weakness (primarily *serratus anterior*) causing muscular instability of the shoulder girdle if axillary nodes are removed

Subcutaneous mastectomy: is a type of skin-sparing mastectomy which removes tissue through an incision under the breast, leaving the skin, areola, and nipple intact. Some women who have prophylactic mastectomies prefer a subcutaneous procedure because it retains their nipples and offers very good cosmetic results. By working through the incision under the breast, the new breast is reconstructed without visible scars. Because a subcutaneous mastectomy leaves more tissue behind—working through the incision under the breast makes it impossible to remove as much tissue as a simple or modified radical mastectomy—this procedure is considered appropriate only as a prophylactic measure. Most physicians consider subcutaneous mastectomy inappropriate for women with large tumors, cancer of the breast skin, or with tumors under or near the nipple or areola. The subcutaneous procedure is different from a 'nipple-sparing mastectomy," where there nipple is scraped free of breast tissue and replaced as a graft.

Potential side effects of subcutaneous mastectomy:

- Skin tightness / adhesions
- Painful and difficult movement of the arm and shoulder

- Increased risk of lymphedema if lymph nodes are removed
- Muscular weakness (primarily *serratus anterior*) causing muscular instability of the shoulder girdle if axillary nodes are removed

Modified radical mastectomy – is the surgical removal of the breast, the nipple, many of the axillary lymph nodes, and the lining over the chest muscle. This procedure has replaced the radical mastectomy as the most common surgery for breast cancer. The nipple and areola may be reconstructed later. The new nipple won't have any sensation, and there will most likely be significant numbness in the remaining skin of the breast.

Potential side effects of modified radical mastectomy:

- Skin tightness / adhesions
- Muscular weakness (primarily *serratus anterior*) causing muscular instability of the shoulder girdle
- Painful and difficult movement of the arm and shoulder
- Increased risk of lymphedema
- Frozen shoulder

Nipple-sparing mastectomy – in this procedure the nipple and areola are left in place while the breast tissue under them is removed. Women who have a small early stage cancer near the outer part of the breast, with no signs of cancer in the skin or near the nipple, are better candidates for nipple-sparing surgery. Cancers that are larger or nearby may mean that cancer cells are hidden in the nipple. Some doctors give the nipple tissue a dose of radiation during or after surgery to try and reduce the risk of the cancer coming back.

There are still some problems with nipple-sparing surgeries. Afterward, the nipple does not have a good blood supply, so sometimes it can wither away or become deformed. Because the nerves are also cut, there is little or no sensation left in the nipple. In some cases, the nipple may look out of place later, mostly in women with large breasts. This type of surgery is not yet widely available.

Potential side effects of nipple-sparing mastectomy:

- Skin tightness / adhesions
- Muscular weakness (primarily *serratus anterior*) causing muscular instability of the shoulder girdle if an axillary lymph node dissection is performed

- Painful and difficult movement of the arm and shoulder
- Increased risk of lymphedema with radiation and axillary lymph node dissection
- Frozen shoulder

Radical mastectomy – surgical removal of the breast, the pectoralis major and minor, all of the axillary lymph nodes, and some additional fat and skin.

Potential side effects of radical mastectomy:

- Deformity / large depression in the chest wall
- Inability to bring arm across the chest in a raised position (*horizontal adduction*)
- Muscular weakness (primarily *serratus anterior*) causing muscular instability of the shoulder girdle
- Reduced shoulder stabilization and ability to rotate the shoulder blade upward, limiting the ability to raise the arm out, away from the body (*abduction*), or in front of the body (*flexion*)
- Possible pulmonary problems
- Increased risk of lymphedema
- Frozen shoulder

Axillary node dissection – a standard dissection consists of sampling the level I (beneath the armpit) and level II nodes (in the armpit itself). Generally about fifteen nodes are taken through a separate incision in the fold of the armpit. The number of nodes is not set in stone. The actual number may vary from as few as four to as many as thirty. If the nodes test negative for cancer, the odds that the disease has infiltrated level III (in front of the shoulder) are less than 1%. In a complete axillary dissection, removing all three levels, usually excises over two dozen nodes.

Potential side effects of axillary node dissection:

- Lymphedema
- Reduced arm and shoulder function (strength and range of motion)
- Weakness in the serratus anterior
- Tightness in the skin under the arm
- Numbness

- Recurrent infections
- Frozen shoulder
- Axillary web syndrome, or lymphatic cording is a visible web of axillary skin overlying palpable cords of tissue. Occurs exclusively in conjuction with ALND. Incidence of 6% between 1-6 weeks post-op. It may be attributed to lymph venous injury during ALND due to either tissue retraction and/or patient positioning. Another theory is that because of the interruption in axillary flow, lymphatic vessels are dialating and scalloping to the skin. Skin dimpling may be present. It may occur not only in the axilla but across the elbow joint, wrist, and/or trunk. Cording can cause decrease in AROM, PROM and decrease function. Pain is usually associated with this. If this is noted treatment is usually that of long tissue stretching and myofascial relief techniques, usually done by a Physical Therapist. It is important that these patients be on a home exercise program of stretching exercises. With treatment, AROM and PROM will improve, pain will decrease but visible signs of cording may always be present.

Sentinel node biopsy – The sentinel node is the first node to receive the lymph that drains from the cancerous area. It will be the node most likely to contain cancerous cells if they have spread from the primary site. If this node is found to be clear, it is thought the nodes beyond it will also be clear of the disease. The procedure is performed by injecting a blue dye into the tumor site. A small incision is made under the armpit and the surgeon identifies and removes the first lymph node to turn blue – the sentinel node. If the sentinel node comes back from pathology testing positive for cancer, the surgeon will usually go on to perform a complete axillary node dissection.

Potential side effects of sentinel node biopsy:

- Minimal tightness in the skin under the arm
- Lymphedema

External beam radiation – uses high-dose x-rays to shrink tumors and kill cancer cells that may have been left behind after spreading away from a tumor that was removed by surgery. These remaining cells are often referred to as *hidden cells*, because they are not detected at the time of surgery. The procedure kills cancerous cells by permanently damaging their DNA in a way that causes them to lose their ability to function and then die. Radiation is usually started three to four weeks after surgery. Typically treatment will be administered five days a week for six to seven weeks. Each treatment takes about thirty minutes.

Potential side effects of external beam radiation:

- Fatigue
- Lymphedema
- Mild to moderate skin irritation
- Severe blistering
- Breast swelling, discomfort, or pain
- Decreased breast size
- Dry cough within three months of treatment
- Increased risk of nerve injury
- Increased risk of lymphedema
- Hardening of the surgical scar (fibrosis)
- Scarring of the heart or lungs (very uncommon)
- Rib fractures

SLNB = sentinel lymph node biopsy
ALND = axillary lymph node dissection
XRT = radiation therapy

Number of lymph nodes removed correlates with the risk of developing lymphedema.
Risk is approximately ½ with SLNB
SLNB = 8%
SLNB + XRT = 17%
ALND =15%
ALND +XRT =30%

Amer et al. Lymphology 2004

Hormonal therapy – is most often used to treat women with advance metastatic breast cancer or as an adjuvant treatment – a therapy that seeks to prevent the recurrence of cancer – for women diagnosed with early stage estrogen-receptor-positive cancer. Estrogen-receptor-positive cancer means that estrogen or progesterone might encourage the growth of breast cancer cells in the body. Normally, estrogen and progesterone bind to certain sites in the breast and in other parts of the body. But during treatment, a hormonal medication binds to these sites instead and prevents estrogen from reaching them. This may help destroy cancer cells that have spread or reduce the chances that your cancer will recur.

Types of drugs (see page 186 for side-effects and warnings)

ERD's - estrogen receptor downregulators :

- The drug is approved for treating hormone-receptor-positive metastatic breast cancer in post-menopausal women with cancer that is no longer responding to hormonal therapy such as Tamoxifen.
- ERDs work by attaching to the hormone receptors on breast cancer cells, blocking them, and causing them to break down and stop working.
- In addition to binding to and blocking estrogen receptors, ERDs also stop or slow down the growth of breast cancer cells by breaking down the receptors.
- With fewer hormone receptors available, fewer cells receive the signal telling them to grow, and the overgrowth of cancer cells can be slowed or stopped.
- Because FASLODEX is administered intramuscularly, it should not be used in patients with certain blood disorders or in patients receiving anticoagulants (blood thinners)

SERMs - selective estrogen receptor modulators:

- Block the actions of estrogen in breast tissues and certain other tissues by "occupying" the estrogen receptors on cells.
- The SERM fits in the estrogen receptor, but it does NOT send messages to the cell nucleus to grow and divide.
- SERMs do send estrogen-like signals when they land in receptors' bone cells, liver cells, and elsewhere in the body.
- This means that SERMs seem to help prevent or slow osteoporosis in post-menopausal women and may help lower cholesterol.

- This dual effect—blocking estrogen in some places and imitating estrogen in other places—allows SERMs to have multiple beneficial effects in many women with breast cancer.

Aromatase Inhibitors:

- Lower the amount of estrogen being produced by the body.
- This method contrasts with that of SERMs or ERDs, which block estrogen's ability to "turn on" cancer cells.
- Limiting the amount of estrogen produced means there is less estrogen available to reach cancer cells and make them grow.
- In post-menopausal women, estrogen is no longer produced by the ovaries, but is converted from androgen, another hormone.
- Aromatase inhibitors keep androgen from being converted to estrogen. That means less estrogen in the bloodstream and less estrogen reaching estrogen receptors to trigger trouble.
- Aromatase inhibitors are used primarily for post-menopausal women with metastatic breast cancer. They cannot stop the ovaries of premenopausal women from producing estrogen; for this reason they can only be used in postmenopausal women.
- In the past, these medications were most commonly used by women who may have already tried other anti-estrogen therapies, such as Tamoxifen, and whose cancer was no longer controlled by those drugs.
- Now with the results of new studies, many doctors recommend an aromatase inhibitor before Tamoxifen for post-menopausal women with metastatic disease.

Hyperthermia – heats the tumor (*local hyperthermia*), an organ or limb (*regional hyperthermia*), or the entire body (*whole-body hyperthermia*) to between 40 and 43 degrees Celsius, making them more susceptible to radiation therapy and chemotherapy.

Potential side effects of hyperthermia:

- Discomfort or even significant local pain in about half the patients treated.

- It can also cause blisters, which generally heal rapidly.
- Less commonly, it can cause burns.

Chemotherapy – uses anticancer agents to systemically destroy cancerous cells. It may also be used to shrink tumors, either to reduce pain, or to make surgery easier. Chemotherapy is a systemic treatment that travels through the bloodstream and kills cancerous cells anywhere in the body by interfering with cell growth and division.

***Potential side effects of chemotherapy*:**

- Hair loss
- Dry skin or rash
- Weight gain/loss
- Immunocompromization
- Fatigue
- Nausea, vomiting, or diarrhea
- Decreased appetite
- Nerve damage causing arm or leg tingling and numbness
- Muscular weakness / fatigue
- Damage to the circulatory system and heart muscle (cardiomyopathy or accelerated atherosclerosis)
- Diabetes and osteoporosis
- Instant menopause – menopausal symptoms
- Peripheral neuropathy of the feet and hands

Chemotherapy port: *A port infusion uses an under-the-skin (subcutaneous) port that has been implanted by a surgeon. The port is located either in the arm or chest, and is connected by a soft, slim catheter tube that goes through your vein all the way to the heart. This catheter protects the vein during treatment. The port is an entry point that the infusion nurse can find each time the patient comes for a treatment, and it can be used for a blood draw, as well as infusion of drugs. The chemotherapy nurse will use a special type of needle to access the port, and won't have to hunt for a good vein to use.*

Immunotherapy –typically referred to as biological response modifiers, or biological therapies, interferons, cytokines, monoclonal antibodies, and vaccine therapies have been used to stimulate the body's immune response. Immunotherapy attempts to overcome the body's tolerance and get it to reject the cancer just as it would a transplanted organ. Under normal conditions our bodies exterminate cancer cells before they can evolve into a tumor. Usually the disease manages to take hold during times when our immunity is down. Typically, for breast cancer, monoclonal antibodies (*MOABs*) are used. They are injected into the patient and then latch onto their assigned antigens, signaling white blood cells to swarm over the cancer cells and annihilate them. *Herceptin* goes after the HER-2 gene found in about one-third of breast cancer patients. Because the MOABs target specific proteins instead of attacking all rapidly growing cells like chemotherapy does, their side effects are relatively minor by comparison.

Potential side effects of immunotherapy:

- Flu-like symptoms for a short time

Bone-Marrow Transplantation (BMT) or Peripheral Blood Stem Cell Transplant (PBSCT) – the goal of the transplant when a patient has cancer is to allow them to undergo high-dose chemotherapy and sometimes radiation as well, which will aggressively attack the cancer cells, and then replace the damaged cells with the healthy ones. Prior to the BMT or PBSCT, a catheter (also known as a "Hickman" or central venous line) will be inserted through a vein in the chest. This tube allows for fluids, nutrition solutions, antibiotics, chemotherapy, or blood products to be delivered directly into the bloodstream without having to repeatedly having to insert a needle into the vein. The catheter can also be used to collect blood samples. In a successful transplant, the new bone marrow migrates to the cavities of the large bone and engrafts and begins producing normal blood cells. If bone marrow or stem cells from a donor is used, it is called allogeneic. If it is not a good genetic match, it will perceive the patient's body as foreign material to be attacked and destroyed. This condition is known as graft vs. host disease. On the other hand, the patient's immune system may destroy the new bone marrow or stem cells. This is known as graft rejection. In certain circumstances, patients may be their own donors. This is called an autologous bone marrow transplant. This procedure is used when the cancer is in the bone marrow is in remission, or if the condition being treated does not involve the bone marrow (breast and ovarian cancer, Hodgkin's disease, Non-Hodgkin's lymphoma, and brain tumors).

The two or three weeks immediately following the procedure are the most critical. Because the high dose chemotherapy and radiation have destroyed the patient's bone marrow and stem cells, disabling the immune system. The patient will be very susceptible to infection and excessive bleeding. Antibiotics and blood transfusions will be given to the patient to help prevent and fight infection. Allogeneic patients will need additional medications to prevent and control graft vs. host disease. Precautions will be taken to minimize the patient's exposure to bacteria and viruses. As the transplant begins to engraft, the patient will gradually be taken off of antibiotics and transfusions. Once the patient is producing a sufficient amount of red and white blood cells and platelets, the patient will be discharged (providing that no other complications have arisen). The typical amount of time spent in the hospital is four to eight weeks. Patients usually cannot return to full-time work or resume normal activities for six months. Contact with the general public is restricted during those six months.

Potential side effects of BMT:

- Immunocomprimization
- Severe flu-like symptoms
- Infection
- Mouth sores
- Temporary confusion
- Bleeding
- Liver disease
- Graft vs. host disease
- Graft failure

Correcting range of motion limitations following surgery

Following breast surgery/reconstruction, shoulder range of motion limitations (ROM) are not uncommon. It is important to address these issues because they can lead to additional joint deterioration and/or frozen shoulder if not corrected. Conducting a ROM assessment with a goniometer will help determine which areas need attention. Before beginning a resistance training program, your client should have 90% or better of the lower end of the ROM norm. Remember that ROM measurements taken while standing are ***typically*** a reflection of strength while measurements taken lying down are ***typically*** a reflection of limited flexibility. This process is described in detail and norms for each plane of motion are listed in the shoulder ROM section of this book. For example, if norms for flexion lying down are 150-180, your client should be at no less than 135 degrees before they do any resistance training in flexion. If they are at 150 degrees standing, that usually insinuates that it is a flexibility issue, not a strength issue. Flexibility limitations ***always*** over rule strength limitations. You ***do not*** want a client to become stronger in a limited plane of motion. Clients may begin resistance training in other planes of motion if they 90% or better of normal range. Therefore your client, who only has 125 degrees of shoulder ROM in flexion, should not do a pullover with any type of resistance, but they can do exercises with resistance in the other planes of motion. It is important for you to consider which exercises take place in any given plane of motion. You can use this determination to make recommendations based on the need for improved ROM or strength. Below you will find some examples of exercises to help correct ROM limitations for strength and flexibility. You can use any exercise you would like - with strength vs. flexibility in mind. They can be conventional strength training, Yoga, Pilates, water-based etc… Many of these exercises are represented with descriptions and photographs in "Essential Exercises for Breast Cancer Survivors" available on our website. In addition, you will find recommended protocol for Pilates, Yoga, and Personal Trainers in the following section. Keep in mind that these corrective exercises apply to ***anyone*** (not just breast cancer patients) who has a specific ROM deficiency.

Range of motion limitations following breast surgery – flexibility

	Pilates	**Personal Trainer**
Shoulder Flexion	**Rib cage arms** **Arm circles** **Small arm circles (1 lb. wgts.)**	**Pullover w/dowel** **Forward wall walk**
Shoulder Extension	**Chest expansion** **Magic Circle at the back** **Small arm circles** (1 lb. wgts)	Bi-lateral - **Shoulder extension** (w/dowel)
Shoulder Abduction	**Small arm circles** (1 lb. wgts) **Saw**	**Side wall walk** **Self-assisted abduction** (hold dowel underhand with unaffected arm and overhand with affected arm. Affected arm should abduct as far as

		possible without assistance then push a little farther with unaffected arm)
Shoulder Internal Rotation	**Band at side internal rotation** **Small arm circles** (1 lb. wgts)	**Back scratcher** (w/dowel)
Shoulder External Rotation	**Band at side external rotation** **Small arm circles** (1 lb. wgts) **Single arm salute** **Palms up/down**	**Back scratcher** (w/dowel) **Traffic cop** (lie on back and begin w/arms at 90° and perpendicular to floor. Externally rotate towards floor)

Range of motion limitations following breast surgery – strength training

	Pilates	**Personal Trainer**
Shoulder Flexion	**Straight down** **Overhead w/Magic Circle**	**Pullover** **Frontal raise** (both w/resistance)
Shoulder Extension	**Chest expansion** **Magic Circle at the back** **Back rowing 45°** **Back rowing sternum**	**Arm extension with bands** **Tricep kick-backs** **Straight arm press down** (all w/resistance)
Shoulder Abduction	**Side-arm series** **Shoulder abduction 45°**	**Snow angels against wall** **Lateral raise**

Shoulder Internal Rotation	**Side-arm series** **Shoulder internal rotation**	**IR** (w/resistance)
Shoulder External Rotation	**Back rowing 45°** **Triceps** **Side-arm series**	**ER** (w/resistance)

Yoga (will have the benefit of strengthening, stabilization, and ROM):

Have your client begin with a breathing exercise-

Corpse Pose (Savasana): this exercise, if done correctly, will stimulate blood circulation and will lessen or relieve fatigue, nervousness, asthma, constipation, diabetes, indigestion, and insomnia. It will also improve one's mental concentration. Breathing should take place through the nose, from the belly, using full capacity of the lungs. Have your client focus on their diaphragmatic breathing, letting the exhaling take a little longer than the inhaling. Have them hold the pose for several minutes, keeping their mind still and focusing on their breathing and their body.
Option: knees bent or place a block under each knee (good suggestion for someone with low back pain)

Shoulder flexion

- *From **Corpse Pose**,* have client bring their elbows to a 90° angle. If necessary, they can rest the back of each of their hands under a block (if the stretch to the floor is too uncomfortable). Have them breathe through the nose from the belly and continue to focus on their diaphragmatic breathing.

- Next have them bring their knees to their chest, gently rocking right and left keeping their hips on the floor.

- From there, have them come onto ***all 4's*** (on hands and knees bringing the wrists underneath the shoulders and the knees underneath the hips) and into ***Extended Child Pose.***
  Options: (1) Let the chest float between the knees (this is a great suggestion after a mastectomy when there may be a great deal of discomfort having the knees pressing against the chest).

  (2) Extend arms straight in front (flexion) or, if they still have limited ROM and they are in too much pain, have them stack their fists and rest their forehead on them until they are comfortable extending their arms.

From the ***Extended Child Pose***, come back onto ***all 4's*** (on hands and knees bringing the wrists underneath the shoulders and the knees underneath the hips).

***Cat Cow Stretch*:** have your client curl in on exhale and curl out on inhale. Next, have client move into spinal balance.

- ***Spinal Balance*:** have client assume a "neutral spine" in an "all fours" position then have them extend one leg out (with the option to extend opposite arm in front). If they can do this with relative ease, have them alternate to the other leg and arm. *One move, one breath: inhale and extend, exhale and close.* From ***Spinal Balance***, have your client come up and flow into ***Chair.***

- ***Chair***: have your client raise their arms as high as they can comfortably. Over time they can work on increasing their range of motion. The elevation of the arms in this pose will also help to promote lymph drainage for anyone who is at risk for upper extremity lymphedema.
  Option: keep one hand on thigh and alternate (good suggestion for someone with low back pain).

- ***Downward facing dog***: this should be executed by a client who has close to full ROM in flexion. Because of the additional force exerted onto the affected limb, have them elevate that arm following the exercise, and pump their fist open and closed. This will help to promote lymph drainage. Monitor for signs of swelling.

* Now that your client has warmed-up and "flowed" through the previous poses, encourage them to do any of the following poses that pertain to their range of motion deviations or concerns.

<u>Shoulder extension</u>

- ***Mountain pose*** feet together arms along the body.

- From ***Mountain pose***, interlace the fingers behind back or use a strap, or ***Reverse Namasté.*** If your client has scar tissue/adhesions across their chest wall from radiation or surgery, they may struggle with this pose at first try. They should proceed gingerly and progress in their own comfort zone. If they have undergone a LAT flap reconstruction they may struggle initially do to the inherent weakness after the latissimus muscle has been removed. The smaller shoulder stabilizers will compensate and eventually allow them to perform this pose with greater ease.

Shoulder abduction

- ***Swan dive*** (one move, one breath): have your client sweep their arms out to their sides and upward. Then have them swan dive down, hinging at the hips.
  Option: keep knees slightly bent when flowing (good suggestion for someone with low back pain).
- ***Warrior II pose***
- ***Side Angle*** **and *Extended Side Angle pose***
- ***Triangle and Extended Triangle pose*** (for all 4 of these poses have client stay within their comfort zone; particularly if they have had an axillary node dissection or have scar tissue/adhesions on their chest wall from surgery or radiation)

Shoulder IR

- ***Eagle pose*** (this will be extremely beneficial to a client who has undergone a LAT flap reconstruction due to the scar tissue and possible adhesions at the posterior surgical site)

Shoulder ER

- ***Plank pose:*** with chest expanding and broadening.
- ***Downward Facing Dog*****:** have your client roll their shoulders away from their ears.
- ***Upward Facing Dog:*** **depression and retraction of the scapulae** (all 3 of these poses will be extremely beneficial to a client who has undergone a mastectomy or radiation due to the scar tissue and possible adhesions at the anterior surgical site; will also help to counter round shoulder syndrome. On the flip side, there is a lot of pressure put on the arms which may be too much for some people to handle. Have them ease into it and stay in their comfort zone. Because of the additional force exerted onto the affected limb, have them elevate that arm following the exercise, and pump their fist open and closed. This will help to promote lymph drainage. Monitor for signs of swelling.)

Leukemia

Leukemia is cancer of the tissues that form blood cells causes the bone marrow, the lymph nodes, and spleen to inundate the circulation with ineffective white blood cells which are unable to defend the body against infection and other harmful agents. It also disrupts the marrows' production of the two types of red blood cells; erythrocytes (carry 02) and thrombocytes (platelets). Patients often suffer excessive bleeding and bruising because their platelet level falls dangerously low.

People with leukemia have many treatment options depending on whether they have acute or chronic leukemia, their age, general health, and whether the leukemia cells were found in the cerebrospinal fluid. People with acute leukemia need to be treated immediately. The goal of treatment is to achieve a remission: destroying signs if leukemia in the body and becoming symptom free. Additional therapy may be needed to prevent relapse. This is referred to as consolidation or maintenance therapy. Many people with acute leukemia can be cured. Patients with chronic leukemia who are symptom-free may not need treatment immediately. Your doctor will carefully watch your health (watchful waiting) so that treatment can begin when symptoms arise. Although watchful waiting helps you to avoid the undesirable side-effects of treatment, some people will choose to treat the leukemia right away, avoiding the risk of the leukemia getting worse before treatment is started. When treatment is needed, it often can control the disease and its' symptoms. While people may receive maintenance therapy to keep the cancer in remission, chronic leukemia is seldom cured with chemotherapy. Stem cell transplants will offer the best hope for a cure.

Supportive care will be needed to prevent and/or treat infections, control pain and symptoms, to relieve the side-effects of treatment and help cope with emotions. Because people with leukemia get infections, they may be given antibiotics and other drugs. Vaccines may be given against the flu and pneumonia. Anemia and bleeding are very common in people with leukemia. Blood transfusions of red blood cells and platelets may be needed to treat anemia and reduce the risk of serious bleeding. Leukemia along with chemotherapy can make the mouth sensitive, easily infected, and likely to bleed. This can lead to problems with malnutrition due to the inability for some people to consume certain foods.

Four main types that are divided into two broad categories:

Acute – produce severe symptoms that come on suddenly, but may oftentimes be mistaken for the flu. Organ enlargement from the build-up of rapidly dividing white blood cells in the spleen, nodes, liver, and men's testicles, can be very painful. Symptoms may also include: changes in appetite level and temperament, joint and bone pain, joint tenderness or swelling, paleness, dizziness, weakness, tendency to bruise or bleed easily,

unexplained bleeding, recurrent infections in skin, gums, lung, and urinary tract, tiny red or brown spots on the skin, if CNS is affected: headache, blurred vision, confusion, and unexplained fever.
Forty percent of patients with acute lymphocytic leukemia will develop leukemia of the CNS, which can interfere with motor skills, and cognition, and trigger headaches, and seizures. Average life expectancy of a patient with untreated acute leukemia is less than 2 months.

Chronic – develops very slowly and often gets discovered accidentally during a routine physical. Although the cells are abnormal, some of them will still have the capacity to develop all the way up to mature cells and function somewhat normally early in the disease, but over time they will become more dysfunctional. The average life expectancy, if untreated, is measured in years, not months. Symptoms include a general feeling of poor health, fatigue, lack of energy, fever, loss of appetite, night sweats, enlarged lymph nodes in neck or groin, enlarged spleen, and anemia.

Treatment options:

Chemotherapy – uses anticancer agents to systemically destroy cancerous cells. It may also be used to shrink tumors, either to reduce pain, or to make surgery easier. Chemotherapy is a systemic treatment that travels through the bloodstream and kills cancerous cells anywhere in the body by interfering with cell growth and division. Chemotherapy may be delivered in a number of different ways:

- Orally
- Intravenously
- Port
- Into the cerebrospinal fluid (Intrathecal) – used when leukemia cells are found in the fluid in and around the brain and spinal cord. This method is used to bypass the blood-brain barrier. It can be administered in one of two ways:
    - Directly into the spinal fluid
    - Under the scalp- this is used in children as well as some adults. Chemotherapy is administered through a catheter called an *Ommaya reservoir* . The doctor can inject the drugs into the catheter, avoiding the pain of injection into the spinal fluid.

Potential side effects of chemotherapy:

- Hair loss
- Dry skin or rash
- Immunocompromization
- Fatigue
- Weight gain/loss
- Nausea, vomiting, or diarrhea
- Decreased appetite
- Nerve damage causing arm or leg tingling and numbness
- Muscular weakness / fatigue
- Damage to the circulatory system and heart muscle (cardiomyopathy or accelerated atherosclerosis)
- Diabetes and osteoporosis
- Peripheral neuropathy of the feet and hands

Radiation Therapy – in lieu of surgery, external-beam radiation therapy can reduce the size of the spleen and correct blood disorders associated with leukemia. It may also be used to shrink masses in the lymph nodes or to treat the disease in the central nervous system. Radiation therapy may be used in combination with chemotherapy (radiotherapy). It may be given in two ways; to a specific area of the body where there is a high concentration of leukemia cells, or they may receive total-body radiation (this is usually given before a stem-cell transplant).

Potential side effects of radiation therapy:

- Lymphedema in treated area

Bone-Marrow Transplantation (BMT) or Peripheral Blood Stem Cell Transplant (PBSCT) – the goal of the transplant when a patient has cancer is to allow them to undergo high-dose chemotherapy and sometimes radiation as well, which will aggressively attack the cancer cells, and then replace the damaged cells with the healthy ones. Prior to the BMT or PBSCT, a catheter (also known as a "Hickman" or central venous line) will be inserted through a vein in the chest. This tube allows for fluids, nutrition solutions, antibiotics, chemotherapy, or blood products to be delivered directly into the bloodstream without having to repeatedly having to insert a needle into the vein. The catheter can also be used to collect blood samples. In a successful transplant, the new bone marrow migrates to the cavities of the large bone and engrafts and begins producing normal blood cells. If bone marrow or stem cells from a donor is used, it is called allogeneic. If it is not a good genetic match, it will perceive the patient's body as foreign material to be attacked and destroyed. This condition is known as graft vs. host disease. On the other hand, the patient's immune system may destroy the new bone marrow or stem cells. This is known as graft rejection. In certain circumstances, patients may be their own donors. This is called an autologous bone marrow transplant. This procedure is used when the cancer is in the bone marrow is in remission, or if the condition being treated does not involve the bone marrow (breast and ovarian cancer, Hodgkin's disease, Non-Hodgkin's lymphoma, and brain tumors).

The two or three weeks immediately following the procedure are the most critical. Because the high dose chemotherapy and radiation have destroyed the patient's bone marrow and stem cells, disabling the immune system. The patient will be very susceptible to infection and excessive bleeding. Antibiotics and blood transfusions will be given to the patient to help prevent and fight infection. Allogeneic patients will need additional medications to prevent and control graft vs. host disease. Precautions will be taken to minimize the patient's exposure to bacteria and viruses. As the transplant begins to engraft, the patient will gradually be taken off of antibiotics and transfusions. Once the patient is producing a sufficient amount of red and white blood cells and platelets, the patient will be discharged (providing that no other complications have arisen). The typical amount of time spent in the hospital is four to eight weeks. Patients usually cannot return to full-time work or resume normal activities for six months. Contact with the general public is restricted during those six months.

Potential side effects of BMT:

- Immunocomprimization
- Severe flu-like symptoms
- Infection
- Mouth sores
- Temporary confusion
- Bleeding
- Liver disease

- Graft vs. host disease
- Graft failure

Adoptive Immunotherapy – this may be used as a method of obtaining a long-lasting remission rather than subjecting relapsed patients to a second transplant. The patient is infused with normal white blood cells taken from the marrow donor's blood. Approximately 80% of relapsed patients went into remission following this procedure.

Potential side effects of adoptive immunotherapy:

- Graft-Versus-Host Disease

Immunotherapy (Interferon) – in chronic myeloid leukemia causes a specific genetic defect called the Philadelphia chromosome, which is present in 9/10 patients' bone-marrow-biopsy specimens. Interferon can make the chromosomal abnormality disappear in 15-20% of patients, sometimes for at least ten years.

Potential side effects of immunotherapy:

- Skin irritation that may occur at injection site
- Dizziness
- Depression and emotional changes
- Hair loss
- Headaches
- Muscle aches/weakness
- Fever
- Infertility
- Pins and needles in the hands and feet
- Effects on circulation

Splenectomy – in chronic leukemia the spleen frequently enlarges – sometimes tremendously – as it accumulates leukemic cells, red blood cells, platelets, and transfused blood cells from the bloodstream. If the spleen becomes severely distended, it can press against the stomach, making it difficult to eat, or against the diaphragm, interfering with breathing. A splenectomy may also be used to manage hemolytic anemia or thrombocytopenia, in which the body destroys red blood cells and/or platelet cells, quicker than the bone marrow can produce replacements.

Potential side effects of splenectomy:

- Although the risk of overwhelming infection after splenectomy is substantial in young children (less than 6 years old), this is rare in adults.
- Prior to surgery, each patient should be vaccinated against the common bacteria (germs) that can cause serious infection in patients who have no spleen: pneumococcus, hemophilus influenza and meningococcus.
- Immunocompromization
- Wound infection, post-operative bleeding, pneumonia, etc. may occur but these problems are uncommon.

Lymphomas

Lymphomas are malignancies of the lymphatic system; the network of vessels and organs in charge of defending our bodies against infection. Patients become increasingly vulnerable to infection. Lymphomas and leukemia are classified as cancers of the tissues that form blood cells. People with AIDS are 60 times more likely to develop non-Hodgkin than the general population. Kidney recipients who receive drugs to suppress the immune system are 40-100% more likely to get non Hodgkin's. These cancers often attack the brain and CNS. Lymphomas account for about 3 percent of all cancer cases in the United States.

Types of lymphoma

Non-Hodgkin's forms in tissue other than the nodes and is grouped according to how aggressively it spreads. Compared to Hodgkin's, these malignancies tend to advance rapidly and follow a less predictable course. There are twenty-five main types of Non-Hodgkin's and they are grouped according to how aggressively they spread as well as the tumor cells' shape and size. Cancers are classified as low grade (indolent) or high grade (aggressive). Contiguous lymphomas are those in which the lymph nodes containing cancer are next to each other. Noncontiguous lymphomas are those in which the lymph nodes containing cancer are not next to each other, but are on the same side of the diaphragm. Once the diagnosis is confirmed, doctors need to "stage" the disease. This will help determine the extent of the disease as well as treatment options. Symptoms of Non-Hodgkin's lymphoma include painless swelling in the neck, armpit, or groin, fevers, night sweats, fatigue, weight loss, itching and reddened patches on the skin, nausea, vomiting, and abdominal pain.

- **Stage I** is divided into stage I and stage IE ("E" stands for extranodal and means that the cancer is found in an organ or tissue other than the lymph nodes)
    - **Stage I**: Cancer is found in a single lymph node area.
    - **Stage IE**: Cancer is found in an organ or tissue other than the lymph nodes.
- **Stage II** is divided into stage II and stage IIE ("E" stands for extranodal and means that the cancer is found in an organ or tissue other than the lymph nodes).
    - **Stage II**: Cancer is found in two or more lymph node areas on the same side of the diaphragm (the thin muscle below the lungs that helps breathing and separates the chest from the abdomen).
    - **Stage IIE**: Cancer is found in an organ or tissue other than the lymph nodes and may have spread to one or more lymph nodes on the same side of the diaphragm.
- **Stage III** is divided into stage III, stage IIIE ("E" stands for extranodal and means that the cancer is found in an organ or tissue other than the lymph nodes), stage IIIS ("S" stands for spleen and means that the cancer is found in the spleen), and stage IIIS+E. Stage III: Cancer is found in lymph node areas on both sides of the diaphragm.
    - **Stage IIIE**: Cancer is found in lymph node areas on both sides of the diaphragm and in one area of a nearby organ or tissue other than the lymph nodes.
    - **Stage IIIS**: Cancer is found in lymph node areas on both sides of the diaphragm and in the spleen.
    - **Stage IIIS+E**: Cancer is found in lymph node areas on both sides of the diaphragm, in one area of a nearby organ or tissue, and in the spleen.

- **Stage IV**, the cancer either is found throughout at least one organ or tissue other than the lymph nodes and may be in lymph nodes near this organ or tissue; or has spread throughout one organ or tissue other than the lymph nodes and has spread to lymph nodes far away from that organ.

Hodgkin's typically affects two age groups – 15 to 34 and over 55. It's easily distinguishable from non-Hodgkin's under the microscope by the presence of unique large cells called Reed Sternberg cells. The disease is very treatable because it generally travels very slowly and predictably down the body from the nodes in the neck to the chest, then down to the abdomen and pelvis. Because it advances slowly, three in four patients can be cured. Once the diagnosis is confirmed, doctors need to "stage" the disease. This will help determine the extent of the disease as well as treatment options. The most common symptoms of Hodgkin's lymphoma are painless swelling in the neck, underarm, or groin, caused by enlarged lymph glands, fevers, fatigue, weight loss, itching, and reddened patches on the skin.

- **Stage I** – the cancer is limited to one lymph node region
- **Stage II** – the cancer is in two different lymph node regions, but is limited to a section of the body either above or below the diaphragm
- **Stage III** – when the cancer moves to both above and below the diaphragm, but hasn't spread to other organs
- **Stage IV** – is the most advanced stage of Hodgkin's Disease. It affects not only the lymph nodes, but also other parts of the body, such as the bone marrow or liver.

Treatment options:

Incisional or core biopsy – removal of part of a lymph node

Excisional biopsy – a tissue sample of an enlarged lymph node is needed to make the cancer diagnosis. The doctor will look for the presence of Reed Sternberg cells. Diagnosing Hodgkin's Disease can be difficult because it's possible to mistake other cells for Reed Sternberg cells.

Fine needle aspiration – removal of part of a lymph using a thin needle.

Bone marrow biopsy – the removal of a small piece of bone and bone marrow by inserting a needle into the hipbone or breastbone. A pathologist views both the bone and bone marrow samples under a microscope to look for signs of cancer.

Gallium scan: A procedure to detect areas of the body where cells, such as cancer cells, are dividing rapidly. A very small amount of radioactive material, gallium, is injected into a vein and travels through the bloodstream. The gallium collects in the bones or other tissues (organs) and is detected by a scanner.

Lumbar puncture: A procedure used to collect cerebrospinal fluid from the spinal column. This is done by placing a needle into the spinal column. This procedure is also called an LP or spinal tap.

Chemotherapy – uses anticancer agents to systemically destroy cancerous cells. It may also be used to shrink tumors, either to reduce pain, or to make surgery easier. Chemotherapy is a systemic treatment that travels through the bloodstream and kills cancerous cells anywhere in the body by interfering with cell growth and division.

Potential side effects of chemotherapy:

- Hair loss
- Dry skin or rash
- Weight gain/loss
- Immunocompromization
- Fatigue
- Nausea, vomiting, or diarrhea
- Decreased appetite
- Nerve damage causing arm or leg tingling and numbness
- Muscular weakness / fatigue
- Damage to the circulatory system and heart muscle (cardiomyopathy or accelerated atherosclerosis)
- Diabetes and osteoporosis
- Peripheral neuropathy of the feet and hands

Bone-Marrow Transplantation (BMT) or Peripheral Blood Stem Cell Transplant (PBSCT) – the goal of the transplant when a patient has cancer is to allow them to undergo high-dose chemotherapy and sometimes radiation as well, which will aggressively attack the cancer cells, and then replace the damaged cells with the healthy ones. Prior to the BMT or PBSCT, a catheter (also known as a "Hickman" or central venous line) will be inserted through a vein in the chest. This tube allows for fluids, nutrition solutions, antibiotics, chemotherapy, or blood products to be delivered directly into the bloodstream without having to repeatedly having to insert a needle into the vein. The catheter can also be used to collect blood samples. In a successful transplant, the new bone marrow migrates to the cavities of the large bone and engrafts and begins producing normal blood cells. If bone marrow or stem cells from a donor is used, it is called allogeneic. If it is not a good genetic match, it will perceive the patient's body as foreign material to be attacked and destroyed. This condition is known as graft vs. host disease. On the other hand, the patient's immune system may destroy the new bone marrow or stem cells. This is known as graft rejection. In certain circumstances, patients may be their own donors. This is called an autologous bone marrow transplant. This procedure is used when the cancer is in the bone marrow is in remission, or if the condition being treated does not involve the bone marrow (breast and ovarian cancer, Hodgkin's disease, Non-Hodgkin's lymphoma, and brain tumors).

The two or three weeks immediately following the procedure are the most critical. Because the high dose chemotherapy and radiation have destroyed the patient's bone marrow and stem cells, disabling the immune system. The patient will be very susceptible to infection and excessive bleeding. Antibiotics and blood transfusions will be given to the patient to help prevent and fight infection. Allogeneic patients will need additional medications to prevent and control graft vs. host disease. Precautions will be taken to minimize the patient's exposure to bacteria and viruses. As the transplant begins to engraft, the patient will gradually be taken off of antibiotics and transfusions. Once the patient is producing a sufficient amount of red and white blood cells and platelets, the patient will be discharged (providing that no other complications have arisen). The typical amount of time spent in the hospital is four to eight weeks. Patients usually cannot return to full-time work or resume normal activities for six months. Contact with the general public is restricted during those six months.

Potential side effects of BMT:

- Immunocomprimization
- Severe flu-like symptoms
- Infection
- Mouth sores

- Temporary confusion
- Bleeding
- Liver disease
- Graft vs. host disease
- Graft failure

Intrathecal chemotherapy – chemotherapeutic drugs are injected into the cerebrospinal fluid. The needle is either inserted in the central canal in the lower back, or the patient has a small repository called an *ommaya reservoir* surgically implanted beneath the scalp. The chemo is channeled through a catheter tube to one of the ventricles in the cerebrum. The major drawback is that most of the chemotherapeutic agents are too lethal to be delivered into the cerebrospinal fluid and the fluid is not able to distribute the drugs any deeper than one to two millimeters into the brain.

External beam radiation – achieves cure rates between 80-90%, depending on the stage of Hodgkin's. There are three fields of radiation:

- **Mantle field** – encompassing the neck, armpits, and chest
- **Abdominal field** – encompasses the spleen and the para-aortic nodes in the upper abdomen
- **Pelvic nodes**

Potential side effects of radiation:

- Lymphedema in the irradiated area

Depending on the size and location of the mass, one, two, or all three fields may be irradiated. Radiation to both the mantle and abdominal field is called *subtotal nodal irradiation*, while irradiating the entire trunk is called *total nodal irradiation.* In non-Hodgkin's lymphomas radiotherapy is usually given to only one side of the diaphragm. Typically it will begin with the mantle field. If discovered early, low-grade lymphomas can be cured by radiation alone.

Biological Therapy – also referred to as immunotherapy because it takes advantage of the body's natural immunity against pathogens. They are beneficial as treatment because they offer anticancer effects without many of the undesirable side effects of standard therapies. The following are some of the most promising for treating lymphoma:

- Monoclonal antibodies – made in a laboratory to help one's own immune systems kill the cancerous cells directly.

- Cytokines – are produced naturally by the body to stimulate the cells in the immune system. They can also be produced artificially and administered in large doses that will have a greater effect. Examples include interferons and interleukins. These stimulate the immunes system while also stimulating the growth of blood cells.

- Vaccines – they are not like vaccines for polio and flu; attempting to prevent the disease. These vaccines are designed to stimulate the immune system to mount a specific response against cancer. They also create a "memory" of the cancer so that in the case of a recurrence, the immune system will activate very early, preventing the formation of a new tumor.

Multiple Myeloma

Multiple myeloma is a cancer of plasma cells, white blood cells that manufacture antibodies. These proteins travel through the bloodstream, binding with foreign substances such as bacteria, for the purpose of destroying them. In multiple myeloma, plasma cells reproduce uncontrollably and take over the marrow. Not only will there be too many plasma cells in the marrow, but also an abundance of the unique antibody they produce. It is referred to as the *M protein.* Presence of the M protein in the blood or urine is how multiple myeloma is diagnosed. Typically, multiple myeloma cells collect in the marrow and the outermost layer of the bones, creating multiple tumors. The myeloma cells also react with the cells that help keep the bones strong. There are 2 major kinds of bone cells that normally work together to keep bones healthy and strong. The cells that lay down new bone are called *osteoblasts*. The cells that break down old bone are called *osteoclasts*. Myeloma cells make a substance that tells the osteoclasts to speed up the dissolving of bone. Since the osteoblasts do not get a signal to put down new bone, old bone is broken down without new bone to replace it. This makes the bones weak and they break easily. Therefore, fractured bones are a major problem in people with myeloma. There is a rare related cancer, called *plasmacytoma,* in which the malignant cells accumulate in one bone to form a single mass. This is another type of abnormal plasma cell growth. Rather than multiple tumors in different locations as in multiple myeloma, there is only one tumor; it is called solitary plasmacytomas.

Solitary plasmacytoma usually develops in a bone, where it may be called an *isolated plasmacytoma of bone*. When a plasmacytoma starts in other tissues (such as the lungs or the lining of the sinuses, throat, or other organs), it is called an *extramedullary plasmacytoma*. Solitary plasmacytomas are most often treated with radiation therapy. Sometimes surgery may be used for a single extramedullary plasmacytoma. As long as no other plasmacytomas are found later on, the patient's prognosis is usually very good. However, since many people with a solitary plasmacytoma will develop multiple myeloma, these people are watched closely for signs of this disease

If a patient has smoldering multiple myeloma or precancerous monoclonal gammopathy of undetermined significance (MGUS), the hematologist may recommend no treatment. In the cased of MGUS, they will be monitored closely for signs of progression to myeloma. Patients whose myeloma is not stable will usually require immediate treatment. Initial treatment choices will depend on the severity of the patient's condition and eligibility for a stem-cell transplant. This will be determined by the patient's age and general health. For those who do not qualify for a transplant, they will be given combination chemotherapy with one of three regimens, depending on the

Eighty percent of myeloma cases occur after age 60 and rarely occurs in people under age 50. There is a much higher incidence rate among African Americans. The five year relative survival rate is 35%. Symptoms include bone pain – often in the back or ribs, broken bones, weakness, fatigue, weight loss, nausea, vomiting, constipation, problems with urination, repeated infections and weakness or numbness in the legs.

Complications:

- Multiple myelomas' main effect is on the bones, especially the spine and rib cage. As the cancer cells grow, they stimulate the activities of other cells to eat away at the bone. Skeletal x-rays of patients often display gaping black holes called *lytic lesions*. These holes leave the bones weak and prone to fracture. The first indicators of the disease are often bone pain in the back or ribs, and broken bones. Although multiple myeloma attacks bone, it is not bone cancer because it originates in the plasma cells.

- Because multiple myeloma erodes the bone, calcium is released into the blood and builds up in large amounts causing symptoms of *hypercalcemia;* nausea, fatigue, and thirst. It is important to try and prevent fractures through weight bearing exercise. A cane or walker can be used to provide a wider base of support. Drinking plenty of fluids is also important, since it helps the kidneys to get rid of excess calcium in the blood and prevents problems that occur when calcium collects in the kidneys.

- As malignant plasma cells invade the marrow, white cells, red cells, and platelets are crowded out and can't produce their usual cells. A deficiency in red blood cells brings on symptoms of *anemia*; fatigue, shortness of breath, and lethargy. Too few platelets in the bloodstream are known as *thrombocytopenia* and can lead to excessive bleeding and bruising.

- In multiple myeloma, the myeloma cells crowd out the normal plasma cells, so that the antibodies to fight the infection aren't made. Infections associated with multiple myeloma include *pneumococcal pneumonia, streptococcus, staphylococcus, and shingles (herpes zoster).* In order to prevent infections, patients should not get any vaccines or inoculations with live materials, they should consume plenty of fluids as well as a diet high in calories and proteins, and should get plenty of rest.

- In three out of four patients, a substance called *Bence-Jones proteins* are produces by the plasma cells, clogging the narrow tubules of the kidneys and damaging the organs. Patients sometimes have impaired renal function and have to go on kidney dialysis while they're being treated for the cancer. The kidney damage may be permanent, but is often reversible with treatment, sustaining minimal damage.

- Spinal cord compression is one of the most severe adverse effects of multiple myeloma. Reports indicate that as many as 20% of patients develop spinal cord compression at some point during the course of their disease. Symptoms typically include back pain, weakness or paralysis in the legs, and numbness in the lower extremities. However, depending on the level of involvement, patients may present with upper extremity symptoms. The dysfunction may be reversible, depending on the duration of the cord compression; however, once established, the dysfunction is only rarely fully reversed.

Treatment options:

Chemotherapy – while multiple myeloma is incurable, patients can live with the disease for many years. Since 1970, the standard chemotherapy regimen has consisted of Alkeran and the steroid prednisone (Deltasone) given for a maximum of a year followed by a rest period. Often chemotherapy is discontinued during what is called a plateau phase of remission, during which the M protein level remains stable. Patients may need chemotherapy again if their M protein levels begin to rise. The major drawback of Alkeran is that it can permanently injure the normal stem cell found in bone marrow. To replenish the marrow, stem cells can be harvested from the patient's marrow and/or skimmed from the bloodstream prior to high dose chemotherapy, then reinfused afterward.

External beam radiation – Radiation uses high-dose x-rays to shrink tumors and kill cancer cells that may have been left behind after spreading away from a tumor that was removed by surgery. These remaining cells are often referred to as *hidden cells*, because they are not detected at the time of surgery. The procedure kills cancerous cells by permanently damaging their DNA in a way that causes them to lose their ability to function and then die. Radiation is usually started three to four weeks after surgery. Typically treatment will be administered five days a week for six to seven weeks. Each treatment takes about thirty minutes.

Potential side effects of external beam radiation:

- Fatigue
- Mild to moderate skin irritation
- Severe blistering
- Dry cough within three months of treatment
- Increased risk of nerve injury
- Increased risk of lymphedema
- Scarring of the heart or lungs (very uncommon)
- Rib fractures

Stem cell transplant - is similar to bone marrow transplant except the cells are collected from stem cells in the bone marrow and then they are frozen. The cells used for transplant can be your own healthy cells (autologous transplant), or they can be collected from a compatible donor (allogeneic transplant). Physicians harvest bone marrow from the pelvic bones or hip in an operating room while the donor is under general anesthesia. The physician inserts a hollow needle into the rear and sometimes the front hipbone, both of which contain a large quantity of bone marrow. The breastbone is another accessible site that is rich in marrow, but this is very rarely used for harvest. The physician often must pierce the bone in several spots to obtain enough marrow for a transplant. The donor will not need stitches but will have some pain and tenderness at the site of the harvest for about a week. The patient then receives a high dose of chemotherapy, which destroys tumor cells but also destroys most or all of the stem cells in the patient's bone marrow. The harvested stem cells are then administered, or transplanted, to help regenerate the patient's blood and immune systems. This procedure is used more frequently than bone marrow transplant because of shortened recovery times and possible decreased risk of infection. Compared to the month long hospital stay following an autologous BMT, there is usually a 10 day hospital stay and an additional 10-16 days for platelet and white blood cell counts to reach desirable ranges.

Potential side effects of peripheral stem cell transplant:

- Graft rejection
- Graft failure
- Infection
- Liver or heart problems
- Shortness of breath
- Garlic smell or taste due to the solution used to freeze stem cells
- Graft-versus-host disease

Bone Marrow Transplantation – having too few stem cells due to the toxicity of Alkeran, puts autologous transplantation out of reach. Administering heavy doses of chemotherapy, then infusing patients with identically matching (allogeneic) marrow stem cells from a sibling, may offer hope of a cure. The outcome is probably better for patients with early stage disease.

Potential side effects of BMT:

- Graft rejection
- Graft failure
- Infection
- Graft-versus-host disease (GVHD)
- Immunocompromization

Interferon - is a hormone-like substance released by some white blood cells and bone marrow cells and it slows the growth of myeloma cells. It is sometimes given to patients who have been treated with chemotherapy so that the myeloma is in remission. Interferon seems to prolong remission. Interferon causes side effects that include fatigue and other symptoms similar to those from a flu infection.

Bortezomib (Velcade®) - is the first in a new class of medicines called proteasome inhibitors, and the first treatment in more than a decade to be approved by the Food and Drug Administration for patients with multiple myeloma. Proteasomes are protein complexes found within all cells that break down proteins and are essential for normal cellular processes like the cell cycle, signal transduction, and gene expression. Inhibiting proteasome activity in cancer cells seems to increase programmed cell death (a process called apoptosis), block proliferation, and inhibit the cell cycle, causing cancer cell death. Currently bortezomib is approved for patients with resistant or relapsed myeloma. Through clinical trials, our investigators are combining bortezomib with other agents as an initial treatment for patients with advanced myeloma (ISS stage II and ISS stage III) in order to achieve prompt disease control. More than half of the patients in the trial have achieved complete or near-complete responses, an outcome that makes stem cell transplant more effective because the target population of myeloma cells is smaller.

Corticosteroids - can be used alone or combined with other drugs. Corticosteroids help decrease the nausea and vomiting that may accompany chemotherapy. These drugs have side effects including high blood sugar, increased appetite, and sleeplessness. When used for prolonged periods, corticosteroids will suppress the immune system. This increased the risk of serious infections. These side effects subside after the drug is stopped. The drugs most often used in treating myeloma are dexamethasone and prednisone

Thalidomide - a medicine first used in the late 1950s as a sleeping pill and to combat the nausea some women experience in the first trimester of pregnancy, was found to cause birth defects and was later banned. In recent years, researchers began testing the drug on other diseases, and it was found to be an effective treatment for leprosy. In the late 1990s, cancer researchers discovered that thalidomide could be used to treat myeloma throughout the course of the disease as both first-line and maintenance treatment, and also for those whose disease has relapsed.

Newer versions of thalidomide, such as lenalidomide (Revlimid®), are designed to be more potent and also appear to have fewer side effects than thalidomide. Researchers theorize that thalidomide and lenalidomide act directly and indirectly in myeloma by promoting the death of cancer cells and by inhibiting myeloma cell growth and survival in bone marrow. Doctors use the combination of lenalidomide and low-dose dexamethasone for many newly diagnosed patients with myeloma who have limited organ damage and are candidates for stem cell transplant.

Pamidronate (Aredia®) or **zoledronic acid** (Zometa®) - can slow bone loss and simultaneously help to alleviate bone pain. Bone disease and pain may also be treated with local radiation, and pathologic fractures can be treated through surgical approaches in which surgeons attach metal rods and plates to weight-bearing bones to provide support.

Observation – one in ten myeloma patients will have what is referred to as stable disease. While it is largely immune to chemotherapy, the cancer cells are so inactive that no treatment is necessary until the disease advances.

Hormonal Therapy – Dexamethasone, a steroid like prednisone, can induce partial remissions in patients who did not respond to their initial round of chemotherapy.

Kidney Dialysis - in ***hemodialysis***, a machine called a *dialyzer* carries out the task of filtering blood three times a week. In ***peritoneal dialysis***, the function is performed several times a day, but inside the person's body.

Plasmapheresis – patients are connected to a machine similar to a *dialyzer*, filtering out the excess myeloma antibodies. Plasmapheresis thins the blood and eases the workload of the kidneys and heart.

Bone-resorption inhibitor (Aredia) – many patients receive monthly intravenous infusions of Aredia which binds to the surface of damaged bones and slows excessive resorption, giving the bones a chance to heal and regain their density and strength. The drug reduces bone fractures, spinal-cord compression, or the need for surgery or radiation by nearly half. Most patients are treated once a month at first, but may be able to be treated less often later on, if they are doing well. Treatment with a bisphosphonate helps prevent further bone damage in multiple myeloma patients.

Bisphosphonate treatment does have a rare but serious side effect called osteonecrosis of the jaw. Patients typically will complain of pain to their doctor or dentist and upon examination they find that part of the jaw bone has died. This can lead to an open sore that does not heal and can also lead to loss of teeth in that area. The jaw bone can also become infected. Having jaw surgery or having a tooth removed should be avoided while on bisphosphonate therapy. Patients have a dental check up before starting treatment. That way if there are any dental problems, they can be taken care of before starting the drug. If osteonecrosis does occur, the doctor will stop the bisphosphonate treatment. Good oral hygiene is a must; flossing, brushing, making sure that dentures fit properly, and having regular dental checkups.

Bone and Soft Tissue Cancer

Bone and soft tissue sarcomas are cancers of the structural and connective tissues of the body. They include bone, cartilage, fibrous connective tissue, fat, muscle, and blood vessels. *Primary* bone cancers (those that begin in the bone) are extremely rare – only about 2,000 new cases a year. By contrast, many cancers that begin in other areas of the body – such as breast, prostate, or lung – my spread to the bone and cause *secondary* tumors there. Soft tissue cancers are also fairly uncommon with a total of about 6,000 new cases a year. Symptoms of bone cancer include pain, swelling or mass, stiffness or tenderness in the affected area, loss of bladder or bowel function (if cancer is in pelvic bones or base of spine), and bone fractures. Symptoms of soft tissue sarcomas include enlarging, non-tender swelling or mass and pressure against nearby nerves and muscles. Until recently, amputation was considered the best treatment for bone cancers in the arm or leg. Limb-sparing surgery is now possible in many cases with the use of pre or post-operative chemotherapy.

Types of bone cancer:

Osteosarcoma – is the most common type of bone cancer. It originates in the newly forming tissue of the bone and develops in the long bones of the arms and legs. It contains immature bone cells that destroy and replace normal tissue, weakening the bone. It is usually found in the area of the knee joint, however, it can occur in the arm or back, or less commonly, in any other bone. The main danger is that has a high tendency to spread to distant areas of the body, particularly the lungs. Osteosarcoma typically affects ages 10-25.

Chondrosarcoma – is a rare tumor that grows in the cartilage. It can often cause swollen joints or restrict range of motion. It is often found in the pelvic bone, long bones, scapula and base of ribs and, less frequently, in the bones of the hand, foot, nose, and base of skull. They can remain slow-growing, but when they become aggressive, can metastasize to the lungs and heart. Chondrosarcoma typically affects ages 30-60.

Ewing's sarcoma – is a tumor of the bone which affects children and young adults. It differs from osteosarcoma in that it tends to be found in bones such as ribs rather than the long bones of the arm and leg. It is uncommon before age 5 and after age 30. It may involve any part of the bony skeleton and may extend into the soft tissue around the bones. Fever, chills and weakness, intermittent pain, and swelling are initial symptoms. When the tumor is found outside the bones, it is known as "soft-tissue" or extra-osseous Ewing's sarcoma. For many years Ewing's sarcoma was considered fatal, but with present treatment methods it is highly treatable and in many cases curable. The most curable cases are those found in the lower jaw, skull, face, scapula, vertebra or clavicle, and those below the elbow or knee. Ewing's sarcoma typically affects the same age group as osteosarcoma; 10-25 years of age.

Malignant giant cell tumor – begins in connective tissue of bone marrow. It may weaken the knees or vertebra and cause bone fractures. Malignant giant cell tumors typically affect ages 40-55.

Fibrosarcoma – this is a very rare form of bone cancer which may occur at any age, but is rare in children. Fibrosarcomas may develop in persons who have had radiation or at the site of a past bone fracture. It is also felt that Paget's disease (for unknown reasons, parts of the skeleton become overactive and dismantle simultaneously rebuild themselves at an abnormally fast rate) may be a predisposing factor in the development of fibrosarcoma.

Skip metastasis - is a tumor nodule located in the same bone as the main tumor, but not in continuity with the tumor. It is usually located in the joint adjacent to the main tumor and is most often a high-grade sarcoma.

Types of soft tissue sarcomas:

Fibrosarcoma – is often located deep in the thigh, arms, or trunk and may be large when it is found. Fibrosarcoma usually affects people in their 40's and 50's.

Synovial cell sarcomas (synoviomas) – are most commonly found in the leg between the thigh and knee, but also occur in the hands and feet. It has a tendency to spread to the lymph nodes. Synovial cell sarcoma usually affects ages 20-30.

Rhabdomyosarcoma – is usually found in the thigh, shoulder, and upper arm. It may be large when found. It has a tendency to spread to the lymph nodes. Rhabdomyosarcoma usually affects people in their 40's and 50's.

Liposarcoma – is often found in fatty tissue of the thigh, arms, legs, or trunk. It is most common in middle-aged men.

Leiomyosarcoma – develops in the smooth muscles of the uterus and back part of abdominal cavity. It is very rare.

Hemangiosarcoma – originates in the blood vessels and is found in the arms, legs, and trunk.

Lymphangiosarcoma – usually develops in the lymph vessels of the arms.

Neurofibrosarcoma – develops in the peripheral nerves of the arms, legs, and trunk.

Kaposi's sarcoma – is cancer that is found in the tissues under the skin or mucous membranes. It causes red-brown or purple patches on the skin, most often on the legs. It spreads to other organs in the body such as the lungs, liver, or intestinal tract. In the 1980's many cases were discovered in people with AIDS. In these cases, lesions may be found in the mouth, nose, lymph nodes, GI tract, lungs, liver, and spleen. KS usually spreads more quickly in patients with AIDS. It may produce significant deterioration in the affected organ. About 30% of AIDS patients develop KS.

Treatment options:

Surgical excision – surgical removal of the *primary* or *secondary* tumor with or without a margin of normal tissue.

Conservative surgical excision – surgery to remove the cancer and a margin of normal tissue less than 2 centimeters.

Wide surgical excision – surgery to remove the cancer and a margin of normal tissue extending to 2-3 centimeters in all directions from the tumor site. This procedure usually takes away a significant amount of muscle and may involve resecting nerves that can affect function and leave areas of numbness.

Radical excision – takes out the entire area containing cancer with margins extending approximately 5 centimeters on all sides.

Mohs micrographic surgery - the tumor is cut from the skin in thin layers. During surgery, the edges of the tumor and each layer of tumor removed are viewed through a microscope to check for cancer cells. Layers continue to be removed until no more cancer cells are seen. This type of surgery removes as little normal tissue as possible.

Limb-sparing surgery – administering high-dose radiation as the first step of treatment may shrink an otherwise inoperable soft-tissue sarcoma to a resectable size. The tumor is removed along with a margin of healthy tissue. The removed segment is replaced with a bone graft or metal prosthetic bone. The margin of tissue that has been removed is then replaced with healthy tissue taken from another part of the body. Additional irradiation is given postoperatively. Before taking this approach, the doctor must be confident that this will control the cancer at least as well as amputation and that the preserved limb will be functional.

Amputation – surgery to remove the cancerous part of the body.

Reconstructive surgery – several procedures are available depending on the patients' functional needs:

- **Fusion (arthrodesis)** – may result in a stiff joint, but permits activities such as running and jumping

- **Arthroplasty** – uses a metallic or bone allograft. The implant, however, is an artificial joint that will not tolerate activities such as jogging, racquet sports, or heavy lifting.

- **Intercallary allograft reconstruction** – transplants of bone, tendon, ligaments, and connective tissue are used. Often a cast or brace must be worn for six to twelve months until the allograft is healed to the host bone.

Chemotherapy – uses anticancer agents to systemically destroy cancerous cells. It may also be used to shrink tumors, either to reduce pain, or to make surgery easier. Chemotherapy is a systemic treatment that travels through the bloodstream and kills cancerous cells anywhere in the body by interfering with cell growth and division.

Potential side effects of chemotherapy:

- Hair loss
- Dry skin or rash
- Immunocompromization
- Fatigue
- Weight gain/loss
- Nausea, vomiting, or diarrhea
- Decreased appetite
- Nerve damage causing arm or leg tingling and numbness
- Muscular weakness / fatigue
- Damage to the circulatory system and heart muscle (cardiomyopathy or accelerated atherosclerosis)
- Diabetes and osteoporosis
- Peripheral neuropathy of the feet and hands

External beam radiation – Radiation uses high-dose x-rays to shrink tumors and kill cancer cells that may have been left behind after spreading away from a tumor that was removed by surgery. These remaining cells are often referred to as *hidden cells*, because they are not detected at the time of surgery. The procedure kills cancerous cells by permanently damaging their DNA in a way that causes them to lose their ability to function and then die. Radiation is usually started three to four weeks after surgery. Typically treatment will be administered five days a week for six to seven weeks. Each treatment takes about thirty minutes.

Potential side effects of external beam radiation:

- Fatigue
- Mild to moderate skin irritation
- Severe blistering
- Increased risk of nerve injury
- Increased risk of lymphedema

Brachytherapy - involves the delivery of radiation therapy directly to the tumor site, and for bone cancer treatment it can be administered in two different ways. In one approach, during surgery, after the surgeon removes the tumor, special tubes called catheters are inserted into the tumor bed, which is the site where the tumor originally existed before being surgically removed. After allowing the surgical wound to heal for five to six days, a radiation oncologist inserts radioactive seeds into each of the catheters. The seeds stay in place for several days, delivering a high dose of radiotherapy to the site. When the treatment is completed, both the radioactive seeds and the catheters are removed. A patient may finish the entire course of treatment within ten to 14 days. In certain situations, brachytherapy may be administered for two to three days combined with external-beam radiation that is administered for five weeks.

Arterial embolization - attempts to limit the blood supply to the tumor by obstructing a major existing vessel, starving it to death. Chemotherapeutic agents are injected directly into the artery. The chemotherapeutic agent is prevented from traveling any further, causing a higher concentration of the drug where it is wanted most.

Potential side effects:

- post-embolization pain in the area that was embolized lasting several hours-several days
- flu-like symptoms

Cryosurgery - is sometimes used in addition to surgery for some patients with bone cancer. After a bone tumor is removed, liquid nitrogen is used to freeze the tumor cavity to subzero temperatures, killing microscopic tumor cells and decreasing the chance of tumor recurrence. The frozen bone is stabilized by filling the tumor cavity with bone graft, cement, or rods and screws to prevent fracture.

Skin Cancer

The skin is the body's largest organ. It protects against heat, sunlight, injury, and infection. Skin also helps control body temperature and stores water, fat, and vitamin D. The skin has several layers, but the two main layers are the epidermis (upper or outer layer) and the dermis (lower or inner layer). Skin cancer begins in the epidermis, which is made up of 3 kinds of cells:

- Squamous cells: thin, flat cells that form the top layer of the epidermis.
- Basal cells: round cells under the squamous cells.
- Melanocytes: found in the lower part of the epidermis

Skin cancer can occur anywhere on the body, but it is most common in skin that has been exposed to sunlight, such as the face, neck, hands, and arms. There are several types of cancer that start in the skin. The most common types are basal cell carcinoma and squamous cell carcinoma, which are non-melanoma skin cancers. Basal cell carcinoma grows slowly and rarely spreads, but it can damage nearby tissue. Basal cell cancer usually appears as a pink or white pearly bump or as an irritated patch. The skin may bleed and crust over in a repeating cycle. Squamous cell carcinoma is less common than basal cell, but it can be more dangerous because it grows more quickly and may spread. Squamous cell cancer may appear as a raised pink bump or scaly patch with an open sore in the center. Both basal and squamous cell cancers rarely become life-threatening. Actinic keratosis is a skin condition that sometimes develops into squamous cell carcinoma. The greatest increase has been in melanoma, the most serious and most deadly type of skin cancer. In fact, the percentage of people with melanoma has more than doubled in the last 30 years. Melanoma develops in the cells that produce melanin – the pigment that gives your skin its color. It can also form in your eye and in rare cases in internal organs such as your intestine. Although they make up a small percentage of all skin cancers, melanomas cause the greatest number of deaths. It's because they are more likely than other skin cancers to spread to different parts of the body.

Melanoma of the Eye (Ocular Melanoma) - the eye is the second most common site in the body for melanomas after skin. However, with only 2,500 cases a year, ocular melanomas are relatively rare and require special techniques for diagnosis and treatment. Like melanoma of the skin, ocular melanoma often begins with the emergence of a cluster of melanocytes called a nevus (mole). More than 10 percent of the population will develop a nevus in the eye during their lifetime. Nearly one in 5,000 of these will become cancerous.

Most ocular melanomas occur in the part of the eye known as the uveal tract, the vascular layer that includes the iris (the pigmented cells surrounding the pupil), ciliary body (the ring-shaped muscle that changes the size of the pupil and the shape of the lens when the eye focuses), and choroid (the pigmented layer under the retina). Choroidal melanoma is the most common type of ocular melanoma. Melanoma also may occur in the eyelid, the conjunctiva (the filmy white covering of the eye), and the optic nerve. Most people with ocular melanoma experience no symptoms until the tumor has become large enough to interfere with vision. Untreated, however, ocular melanoma may spread through the bloodstream to other organs. Melanomas of the iris often respond better to treatment than those of the ciliary body.

Treatment options:

Excisional Biopsy- all or part of the abnormal-looking growth is cut from the skin and viewed under a microscope to see if cancer cells are present.

Shave biopsy - a sterile razor blade is used to "shave-off" the abnormal-looking growth.

Punch biopsy - a special instrument called a punch is used to remove a circle of tissue from the abnormal-looking growth.

Mohs micrographic surgery - the tumor is cut from the skin in thin layers. During surgery, the edges of the tumor and each layer of tumor removed are viewed through a microscope to check for cancer cells. Layers continue to be removed until no more cancer cells are seen. This type of surgery removes as little normal tissue as possible and is often used to remove skin cancer on the face.

Electrodessication and curettage - the tumor is cut from the skin with a curette (a sharp, spoon-shaped tool). A needle-shaped electrode is then used to treat the area with an electric current that stops the bleeding and destroys cancer cells that remain around the edge of the wound. The process may be repeated one to three times during the surgery to remove all of the cancer.

Sentinel node biopsy – in this procedure, a radiolabeled material is injected near the tumor site, where lymph fluid will carry it to the "sentinel" node -- the first lymph node to which cancer cells are likely to spread from the primary tumor. The sentinel node is then removed and examined to determine the presence of melanoma cells. Surgery to remove the affected lymph nodes may be necessary if the biopsy indicates metastasis. However, if there is no evidence that cancer has spread, further surgery on the lymph nodes may be avoided. Knowledge of the status of the sentinel node is very important in helping to assess the risk of recurrence after treatment of the initial primary tumor.

Cryosurgery - a treatment that uses an instrument to freeze and destroy abnormal tissue, such as carcinoma in situ. This type of treatment is also called cryotherapy.

Laser surgery - a surgical procedure that uses a laser beam (a narrow beam of intense light) as a knife to make bloodless cuts in tissue or to remove a surface lesion such as a tumor.

Dermabrasion - removal of the top layer of skin using a rotating wheel or small particles to rub away

Chemotherapy – uses anticancer agents to systemically destroy cancerous cells. It may also be used to shrink tumors, either to reduce pain, or to make surgery easier. Chemotherapy is a systemic treatment that travels through the bloodstream and kills cancerous cells anywhere in the body by interfering with cell growth and division.

Potential side effects of chemotherapy:

- Hair loss
- Dry skin or rash
- Immunocompromization
- Fatigue
- Weight gain/loss
- Nausea, vomiting, or diarrhea
- Decreased appetite
- Nerve damage causing arm or leg tingling and numbness
- Muscular weakness / fatigue
- Damage to the circulatory system and heart muscle (cardiomyopathy or accelerated atherosclerosis)
- Diabetes and osteoporosis
- Peripheral neuropathy of the feet and hands

External beam radiation – Radiation uses high-dose x-rays to shrink tumors and kill cancer cells that may have been left behind after spreading away from a tumor that was removed by surgery. These remaining cells are often referred to as *hidden cells*, because they are not detected at the time of surgery. The procedure kills cancerous cells by permanently damaging their DNA in a way that causes them to lose their ability to function and then die. Radiation is usually started three to four weeks after surgery. Typically treatment will be administered five days a week for six to seven weeks. Each treatment takes about thirty minutes.

Potential side effects of external beam radiation:

- Fatigue
- Mild to moderate skin irritation
- Severe blistering
- Increased risk of nerve injury
- Increased risk of lymphedema

Immunotherapy - immunotherapy belongs to a group of proteins known as cytokines. They are produced naturally by white blood cells in the body in response to infection, inflammation, or stimulation. They have been used as a treatment for certain viral diseases, including hepatitis B.

***Potential side effects of immunotherapy*:**

- Flu-like symptoms (magnified at greater doses)

Breast Reconstructive Procedures

Saline implants / tissue expanders – A saline implant is a "balloon" filled with saline and placed beneath the skin either on top of or beneath the chest muscles. A tissue expander is similar, but only partially filled before placement. It, too, goes beneath the skin (either on top of or underneath the chest muscle) initially to stretch the skin. At various times over the course of several weeks, the expander is filled with more fluid. This process is repeated until the expander is full.

Potential side effects:

- Capsular contracture – a condition in which scar tissue around the implant or expander hardens and then contracts. This can cause deformity, pain, and abnormal firmness of the breast.
- Pectoralis major may go into painful spasms
- Rupture
- Loss or changes in nipple and breast sensation

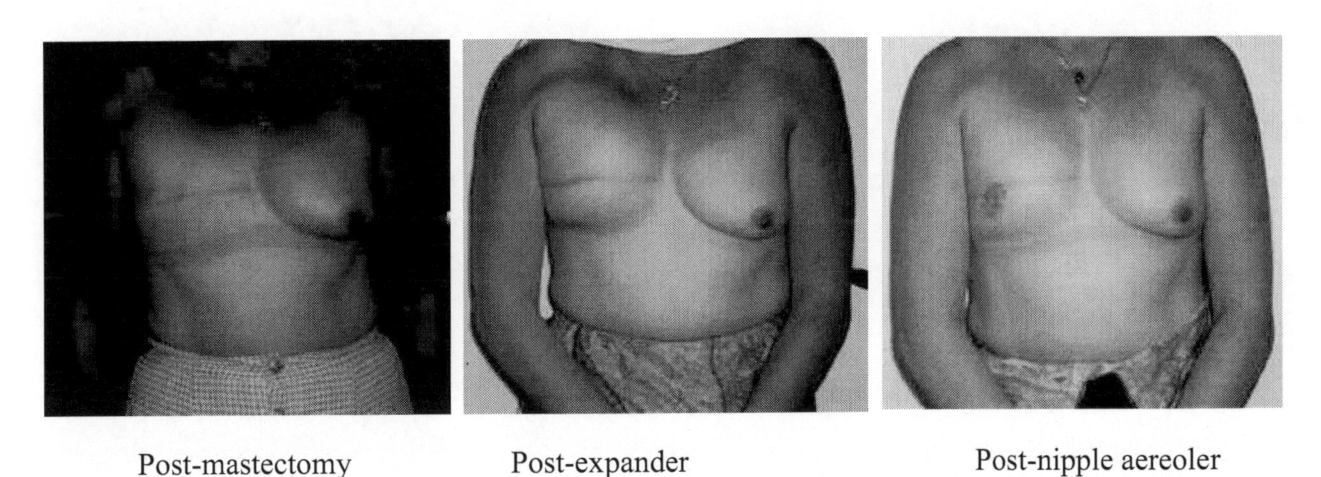

Post-mastectomy | Post-expander | Post-nipple aereoler reconstruction

Latissimus Dorsi Flap – In this procedure, a back muscle called the latissimus dorsi and a football-shaped area of skin are brought to the front of the chest. This is done by creating a "tunnel" under the skin of the armpit and pulling the muscle through the tunnel and out the mastectomy scar in the front. The muscle is then used to form a breast mound, or, more commonly, is folded to create a pocket in which an implant is placed. The skin taken form the back is sewn into place on the front of the chest with all of the blood vessels remaining in tact.

Potential side effects:

- Weakness in the muscles supporting the shoulder blade

- Tissue death (*necrosis*)
- Blood clots
- Infection
- Prolonged healing time
- Loss or changes in nipple and breast sensation

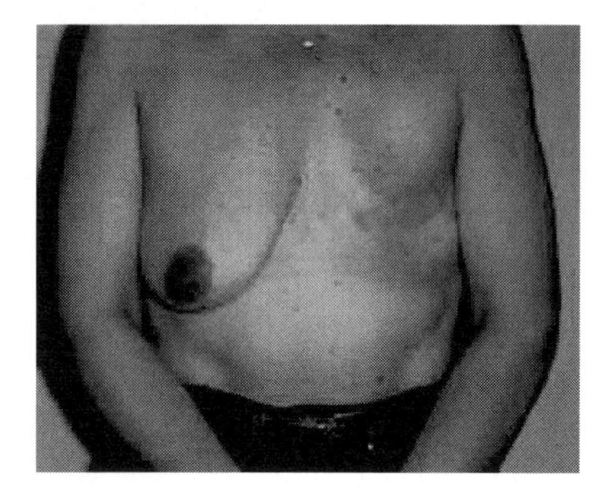

Post-mastectomy

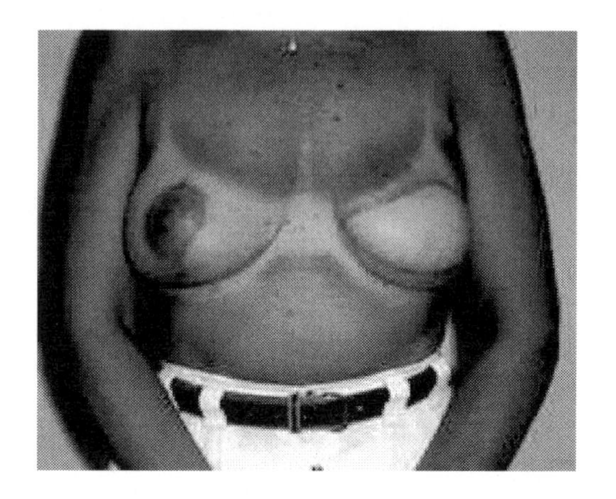

Post-Lat Flap reconstruction

Postural deviations and corrections following a Lat Flap:

Following a Lat Flap procedure, you can expect to encounter the following postural deviations with your clients. Correcting postural deviations should take precedence over any other exercise programming to ensure proper biomechanics and form. Below you will find suggested exercises for correcting various deviations. These are only a sampling of the exercises that you can recommend. ***These corrective exercises can and should be used to correct these specific muscle balances with anyone who has a deviation (any type of cancer, or otherwise healthy population; male or female).***

Round shoulder syndrome or hyperextension of the shoulder – the goal with this client will be primarily to stretch the chest and strengthen the upper back. Typically following a mastectomy/reconstruction, there will be excessive scar tissue and/or adhesions across the chest wall, leading to the rounding forward of the shoulders. This puts excessive strain on the upper back and neck muscles and can lead to a painful imbalance. Focus on stretching/strengthening exercises with the focus on stretching the chest and strengthening the upper back.

Pilates	Personal Trainer
Hug a tree	**Chest fly (no weight)**
Spine stretch forward	**Door or corner stretch**
Chest expansion	**Active Isolated or PNF stretching**
Rowing into sternum	**High/low row**
Rowing at 45°	**Reverse fly**

*** Avoid "pushing" type exercises that may shorten the pectoralis major and exacerbate round shoulder syndrome.**

Yoga (will have the benefit of strengthening, stabilization, and ROM):

Have your client begin with a breathing exercise-

Corpse Pose (Savasana): this exercise, if done correctly, will stimulate blood circulation and will lessen or relieve fatigue, nervousness, asthma, constipation, diabetes, indigestion, and insomnia. It will also improve one's mental concentration. Breathing should take place through the nose, from the belly, using full capacity of the lungs. Have your client focus on their diaphragmatic breathing, letting the exhaling take a little longer than the inhaling. Have them hold the pose for several minutes, keeping their mind still and focusing on their breathing and their body.
Option: knees bent or place a block under each knee (good suggestion for someone with low back pain)

Upper back strengthening

- ***Spinal balance***: have client assume a "neutral spine" in an "all fours" position then have them extend one leg out (with the option to extend opposite arm in front). If they can do this with relative ease, have them alternate to the other leg and arm. *One move, one breath: inhale and extend, exhale and close.*

- ***Locust pose:*** have client ease into pose as they figure out what their ROM limitations are. Because they will most likely have an implant under the LAT flap, advise them to listen to their body and if they feel undue pain or pressure on the chest wall have them try one of the options below.
  Options: feet on or off the floor; hands either under forehead (elbows out); arms along the body; or arms over head.

- ***Sphinx pose***

Chest stretching exercises

- ***Locust pose:*** have client ease into pose as they figure out what their ROM limitations are. Because they will most likely have an implant under the LAT flap, advise them to listen to their body and if they feel undue pain or pressure on the chest wall have them try one of the options below.
  Options: feet on or off the floor; hands either under forehead (elbows out); arms along the body; or arms over head.

- ***Bow pose*** - have client ease into pose as they figure out what their ROM limitations are. Because they will most likely have an implant under the LAT flap, advise them to listen to their body and if they feel undue pain or pressure on the chest wall have them try the ***half bow pose***.
  Option: *Half Bow pose*

- ***Upward Facing Dog:*** this should be executed by a client who has close to full ROM in flexion. Because of the additional force exerted onto the affected limb, have them elevate that arm following the exercise, and pump their fist open and closed. This will help to promote lymph drainage. Monitor for signs of swelling.
- ***Table Top*** (fists for wrists if wrists concerns*):* this is somewhat of an advance move for a client lacking a latissimus muscle and having to contend with tightness and possible scar tissue in the chest wall, however it is very effective. Save this exercise for your client who is further out from their surgery date and has fully recovered, has moderate strength, and good flexibility in shoulder extension.

- ***Staff pose:*** they should proceed gingerly and progress in their own comfort zone. They may struggle initially do to the inherent weakness after the latissimus muscle has been removed. The smaller shoulder stabilizers will compensate and eventually allow them to perform this pose with greater ease. Because of the additional force exerted onto the affected limb, have them elevate that arm following the exercise, and pump their fist open and closed. This will help to promote lymph drainage. Monitor for signs of swelling.
- In all the poses, depression and retraction of the scapulae are imperative for shoulder stabilization.

Shoulder girdle stabilization

- ***Downward Facing Dog****:* have your client roll their shoulders away from their ears.

- ***Dolphin Plank pose*** : shoulders over the elbows

  Option: knees on the floor

- ***Dolphin pose***

* In all poses, depression and retraction of the scapulae are imperative for shoulder stabilization.

"Winged" scapula – following an axillary lymph node dissection, there may be a temporary cease-fire to the serratus anterior muscle on that side. This may present itself as a protraction or "winging" of the scapula. The goal is to perform exercises that will help strengthen the serratus anterior and increase its' ability to stabilize the scapula.

Pilates	Personal Trainer
	Serratus reach – have client lie on back with both arms extended over chest and palms facing each other. Have them raise their shoulders off of floor and reach up as high as they can without lifting their head or back. **Wall push-ups** **Scapular retraction/depression**

Yoga (will have the benefit of strengthening, stabilization, and ROM):

Have your client begin with a breathing exercise-

Corpse Pose (Savasana): this exercise, if done correctly, will stimulate blood circulation and will lessen or relieve fatigue, nervousness, asthma, constipation, diabetes, indigestion, and insomnia. It will also improve one's mental concentration. Breathing should take place through the nose, from the belly, using full capacity of the lungs. Have your client focus on their diaphragmatic breathing, letting the exhaling take a little longer than the inhaling. Have them hold the pose for several minutes, keeping their mind still and focusing on their breathing and their body.

Option: knees bent or place a block under each knee (good suggestion for someone with low back pain)

Shoulder girdle stabilization

- ***Downward Facing Dog***: have your client roll their shoulders away from their ears.
- ***Dolphin Plank pose*** : shoulders over the elbows
  Option: knees on the floor
- ***Dolphin pose***

* In all poses, depression and retraction of the scapulae are imperative for shoulder stabilization.

Kyphosis (forward head) – many of your clients will show signs of slight to moderate forward head, irregardless of whether they have undergone a mastectomy. It may be due to poor posture or improper ergonomics at work. Following a mastectomy, it is often exacerbated by the tightening of the chest wall as well as the "guarded" position many women will take after surgery. This common postural deviation can cause a chain reaction leading to neck and back pain and, if untreated, can lead to a "hunched-back" appearance. In addition to the exercises to stretch the chest wall and strengthen the upper back, the following exercises will help to strengthen the neck muscles, bringing the head back into alignment.

Pilates/Yoga	Personal Trainer
Head retractions	**"Chicken"** – retract head, keeping head in neutral **Isometric retraction** against car seat head rest

Trans Rectus Abdominis Myocutaneous (TRAM) Flap – In this procedure, a football-shaped section of skin, fat, blood vessels, and part, or all of the rectus abdominis muscle from the abdominal area is pulled up through a "tunnel" under the skin of the upper abdomen to form a breast mound on the chest. The skin is sewn into place with all of the nerves and blood vessels remaining intact. This procedure can also be done as a Free Flap (See definition below) for those who may not be candidates for the TRAM flap due to prior surgeries or compromised blood vessels.

Potential side effects:

- Abdominal and lower back weakness and pain

- Hernia
- Decreased trunk stability
- Difficulty standing erect
- Tissue death (*necrosis*)
- Blood clots
- Infection
- Prolonged healing time
- Loss or changes in nipple and breast sensation

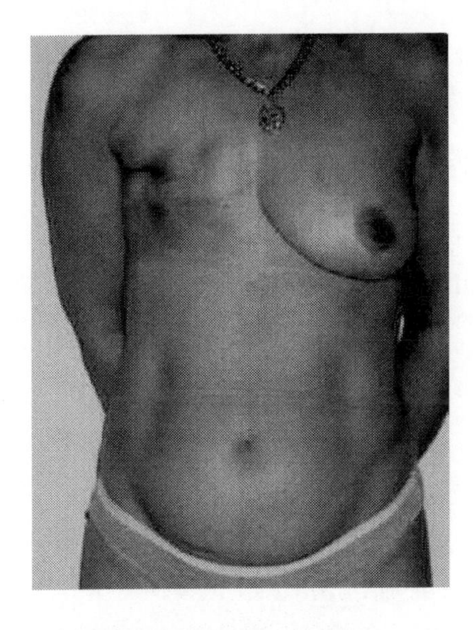

Post-mastectomy

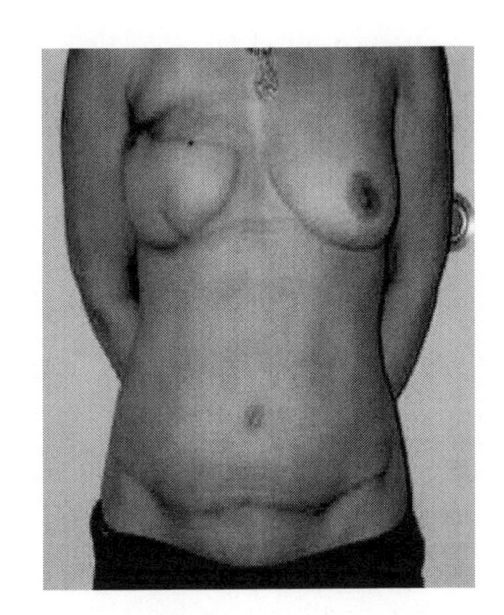

TRAM reconstruction

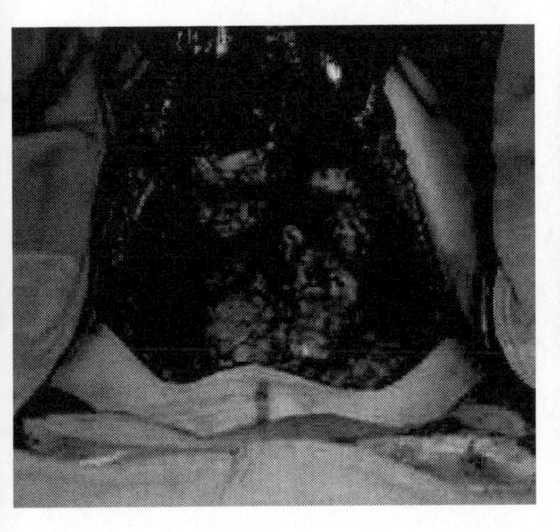

Surgery – Rectus pedicles

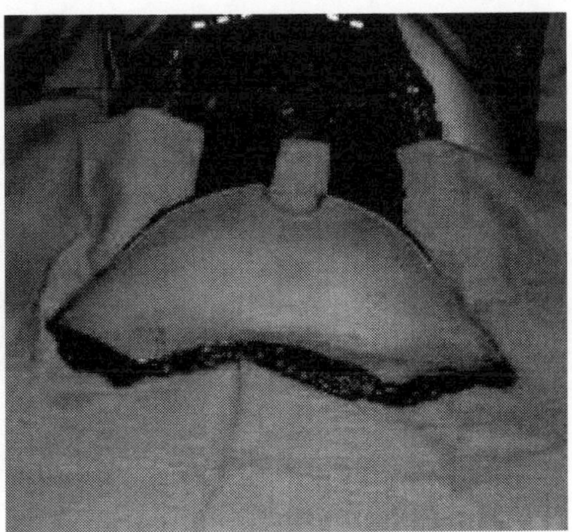

Surgery – Bi Lateral TRAM

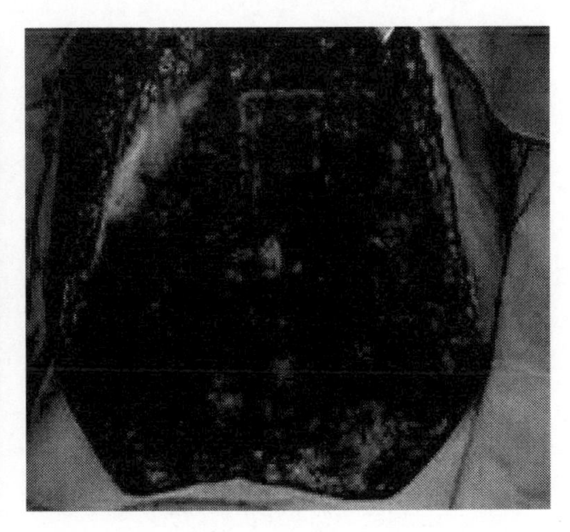

Surgery – marlex mesh

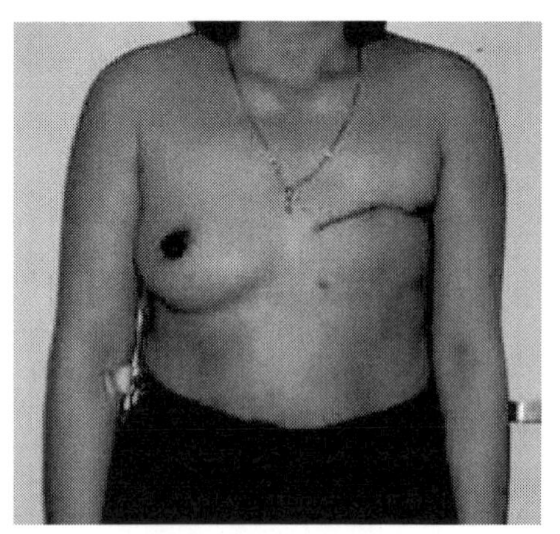

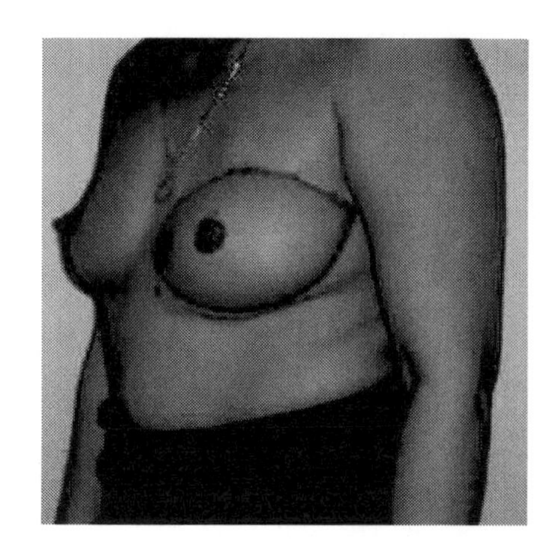

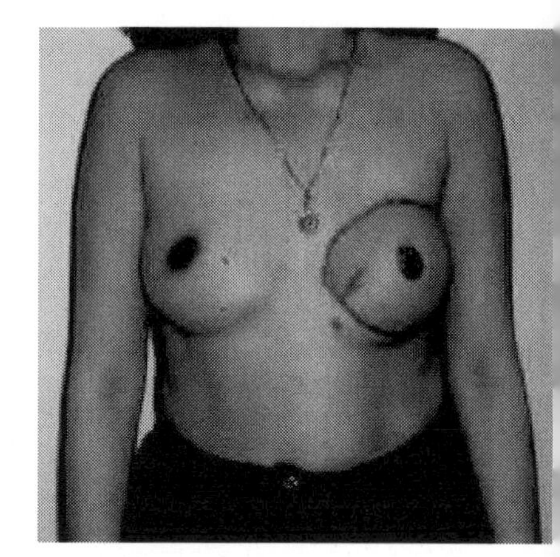

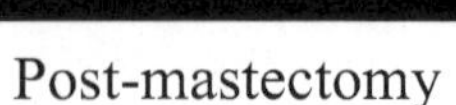

Post-mastectomy

Post TRAM and nipple-aereolar Reconstruction

Postural Deviations and Corrections Following TRAM:

Following a TRAM procedure, you can expect to encounter the following postural deviations with your clients. Correcting postural deviations should take precedence over any other exercise programming to ensure proper biomechanics and form. Below you will find suggested exercises for correcting various deviations. These are only a sampling of the exercises that you can recommend. If you understand the principle behind the chosen exercise, you can adapt it to include any modality; Yoga, Pilates, water exercise, weight training…..

Round shoulder syndrome or hyperextension of the shoulder – the goal with this client will be primarily to stretch the chest and strengthen the upper back. Typically following a mastectomy/reconstruction, there will be excessive scar tissue and/or adhesions across the chest wall, leading to the rounding forward of the shoulders. This puts excessive strain on the upper back and neck muscles and can lead to a painful imbalance. Focus on stretching/strengthening exercises with the focus on stretching the chest and strengthening the upper back. Avoid "pushing" type exercises that may shorten the pectoralis major and exacerbate round shoulder syndrome.

Pilates	Personal Trainer
Hug a tree	**Chest fly (no weight)**
Spine stretch forward	**Door or corner stretch**
Chest expansion	**Active Isolated or PNF stretching**
Rowing into sternum	**High/low row**
Rowing at 45°	**Reverse fly**

Yoga (will have the benefit of strengthening, stabilization, and ROM):

Have your client begin with a breathing exercise-

Corpse Pose (Savasana): this exercise, if done correctly, will stimulate blood circulation and will lessen or relieve fatigue, nervousness, asthma, constipation, diabetes, indigestion, and insomnia. It will also improve one's mental concentration. Breathing should take place through the nose, from the belly, using full capacity of the lungs. Have your client focus on their diaphragmatic breathing, letting the exhaling take a little longer than the inhaling. Have them hold the pose for several minutes, keeping their mind still and focusing on their breathing and their body.

Option: knees bent or place a block under each knee (good suggestion for someone with low back pain)

Upper back strengthening

- ***Spinal balance***: have client assume a "neutral spine" in an "all fours" position then have them extend one leg out (with the option to extend opposite arm in front). If they can do this with relative ease, have them alternate to the other leg and arm. *One move, one breath: inhale and extend, exhale and close.*

- ***Locust pose:*** have client ease into pose as they figure out what their ROM limitations are. Because they will most likely have an implant under the LAT flap, advise them to listen to their body and if they feel undue pain or pressure on the chest wall have them try one of the options below.

  Options: feet on or off the floor; hands either under forehead (elbows out); arms along the body; or arms over head.

- ***Sphinx pose***

Chest stretching

- ***Locust pose:*** have client ease into pose as they figure out what their ROM limitations are. Because they will most likely have an implant under the LAT flap, advise them to listen to their body and if they feel undue pain or pressure on the chest wall have them try one of the options below.

  Options: feet on or off the floor; hands either under forehead (elbows out); arms along the body; or arms over head.

- ***Bow pose*** - have client ease into pose as they figure out what their ROM limitations are. Because they will most likely have an implant under the LAT flap, advise them to listen to their body and if they feel undue pain or pressure on the chest wall have them try the ***half bow pose***.

Option: *Half Bow pose*

- ***Upward Facing Dog:*** this should be executed by a client who has close to full ROM in flexion. Because of the additional force exerted onto the affected limb, have them elevate that arm following the exercise, and pump their fist open and closed. This will help to promote lymph drainage. Monitor for signs of swelling.

- ***Table Top*** (fists for wrists if wrists concerns*):* this is somewhat of an advance move for a client lacking a latissimus muscle and having to contend with tightness and possible scar tissue in the chest wall, however it is very effective. Save this exercise for your client who is further out from their surgery date and has fully recovered, has moderate strength, and good flexibility in shoulder extension.

- ***Staff pose:*** they should proceed gingerly and progress in their own comfort zone. They may struggle initially do to the inherent weakness after the latissimus muscle has been removed. The smaller shoulder stabilizers will compensate and eventually allow them to perform this pose with greater ease. Because of the additional force exerted onto the affected limb, have them elevate that arm following the exercise, and pump their fist open and closed. This will help to promote lymph drainage. Monitor for signs of swelling.

* In all the poses, depression and retraction of the scapulae are imperative for shoulder stabilization.

Shoulder girdle stabilization

- ***Downward Facing Dog****:* have your client roll their shoulders away from their ears.

- ***Dolphin Plank pose*** : shoulders over the elbows
  Option: knees on the floor

- ***Dolphin pose***

* In all poses, depression and retraction of the scapulae are imperative for shoulder stabilization.

"Winged" scapula – following an axillary lymph node dissection, there may be a temporary cease-fire to the serratus anterior muscle on that side. This may present itself as a protraction or "winging" of the scapula. The goal is to perform exercises that will help strengthen the serratus anterior and increase its' ability to stabilize the scapula.

Pilates	Personal Trainer
	Serratus reach – have client lie on back with both arms extended over chest and palms facing each other. Have them raise their shoulders off of floor and reach up as high as they can without lifting their head or back. **Wall push-ups** **Scapular retraction/depression**

Yoga (will have the benefit of strengthening, stabilization, and ROM):

Have your client begin with a breathing exercise-

Corpse Pose (Savasana): this exercise, if done correctly, will stimulate blood circulation and will lessen or relieve fatigue, nervousness, asthma, constipation, diabetes, indigestion, and insomnia. It will also improve one's mental concentration. Breathing should take place through the nose, from the belly, using full capacity of the lungs. Have your client focus on their diaphragmatic breathing, letting the exhaling take a little longer than the inhaling. Have them hold the pose for several minutes, keeping their mind still and focusing on their breathing and their body.
Option: knees bent or place a block under each knee (good suggestion for someone with low back pain)

Shoulder girdle stabilization

- ***Downward Facing Dog****:* have your client roll their shoulders away from their ears.

- ***Dolphin Plank pose*** : shoulders over the elbows
  Option: knees on the floor

- ***Dolphin pose***

* In all poses, depression and retraction of the scapulae are imperative for shoulder stabilization.

Kyphosis (forward head) – many of your clients will show signs of slight to moderate forward head, irregardless of whether they have undergone a mastectomy. It may be due to poor posture or improper ergonomics at work. Following a mastectomy, it is often exacerbated by the tightening of the chest wall as well as the "guarded" position many women will take after surgery. This common postural deviation can cause a chain reaction leading to neck and back pain and, if untreated, can lead to a "hunched-back" appearance. In addition to the exercises to stretch the chest wall and strengthen the upper back, the following exercises will help to strengthen the neck muscles, bringing the head back into alignment.

Pilates/Yoga	**Personal Trainer**
Head retractions	**"Chicken"** **Isometric head against car seat head rest**

Forward flexion – immediately following surgery, and up to several weeks later, your client will have a difficult time standing erect. She will have a large incision across her lower abdomen. There will most likely be pain at the incision site and/or scar tissue that can be a deterrent standing erect. At this stage of recovery, your focus will be to help your client to increase her ROM and flexibility in her torso so she can stand as close to 180 degrees as possible. Be careful not to overstretch, or to put your client in a compromising position that would require them to use their abdominal muscles.

Pilates	Personal Trainer
Bridge on Reformer or mat **Hip lift on Reformer** **Swimming on Reformer** **Knee stretches on Reformer** **Elephant**	**End to end stretch** **Stretching over physioball** **Back extension/Roman Chair** **"Superman"** **Marching on foam roller** **Opposite arm/leg on physioball** **Hip circles/pelvic tilts on physioball** **Stability/balance series on discs**

Yoga (will have the benefit of strengthening, stabilization, and ROM):

Have your client begin with a breathing exercise-

Corpse Pose (Savasana): this exercise, if done correctly, will stimulate blood circulation and will lessen or relieve fatigue, nervousness, asthma, constipation, diabetes, indigestion, and insomnia. It will also improve one's mental concentration. Breathing should take place through the nose, from the belly, using full capacity of the lungs. Have your client focus on their diaphragmatic breathing, letting the exhaling take a little longer than the inhaling. Have them hold the pose for several minutes, keeping their mind still and focusing on their breathing and their body.
Option: knees bent or place a block under each knee (good suggestion for someone with low back pain)

Stretching abdomen

- *Sphinx pose*
- *Camel pose (standing or kneeling)*
- *Bridge pose*

Lordosis – once your client is standing erect, there is a good chance that the area that was once tight and bound down will give way to muscle weakness from the absence of the rectus abdominis. Most people have varying degrees of lordosis to begin with, however, it will be exacerbated by the lack of support in the abdominal wall. The low back may begin to compensate by going into painful spasms. Your client will need to focus on core strengthening exercise while also learning appropriate back stretches and back strengthening exercises.

Pilates	Personal Trainer
Bridge on Reformer or mat	**End to end stretch**
Hip lift on Reformer	**Stretching over physioball**
Swimming on Reformer	**Back extension/Roman Chair**
Knee stretches on Reformer	**"Superman"**
Elephant	**Marching on foam roller**
Hundred	**Opposite arm/leg on physioball**
Ab Prep	**Hip circles/pelvic tilts on physioball**
Long stretch	**Stability/balance series on discs**
	Pelvic tilts/bridging on mat
	Oblique exercises
	"Dying Bug"
	Cable hip extension
	Stretch hip flexors

	Pelvic floor exercises

Yoga (will have the benefit of strengthening, stabilization, and ROM):

Have your client begin with a breathing exercise-

Corpse Pose (Savasana): this exercise, if done correctly, will stimulate blood circulation and will lessen or relieve fatigue, nervousness, asthma, constipation, diabetes, indigestion, and insomnia. It will also improve one's mental concentration. Breathing should take place through the nose, from the belly, using full capacity of the lungs. Have your client focus on their diaphragmatic breathing, letting the exhaling take a little longer than the inhaling. Have them hold the pose for several minutes, keeping their mind still and focusing on their breathing and their body.
Option: knees bent or place a block under each knee (good suggestion for someone with low back pain)

Lower back/hip flexors

- *Bridge pose*
- *Fish pose.* Option for neck concerns: Keep the gaze toward the ceiling.
- *Staff pose*: Option: lean on fingers behind to support back.

Core Strength – Pelvic floor

- *Boat pose*
  Option*: keep knees bent if tight hamstrings*

- *Plank pose* with *Half Crocodile* and back into Plank.
  Option: *kneeling Plank* and *Half Crocodile* on knees.

- In *supine position*, bring hands behind ears, bring knees over hips in 90° angle, feet flexed, hips neutral position. Lower one leg (toes or heel may touch the floor), come back at 90° angle. Keep alternating sides.
  Options: extend legs vertical position and lower one leg at the time keeping hips stable and in neutral position, extended arms over head.

Obliques

- ***Gate pose***
- ***Extended Angle pose***
- ***Extended Triangle pose***
- ***Belly Twist*:** in supine position, bring knees at 90° angle, open arms to the sides shoulder height, let the knees float to one side making sure opposite shoulder does not lift off the floor. Option: Place a block under the knees in order to keep opposite shoulder on the floor.

Free Flap – This procedure is similar to the TRAM flap in that skin, muscle (in a rectus abdominis flap), fat, and blood vessels are transferred from the abdomen, or other area of the body, to the chest to make a breast mound. The difference, however, is that instead of leaving all of the tissue intact and tunneling it under the skin and up to the chest, the tissue – blood vessels and all – is cut out and then reattached to the chest. This is a much more complicated and intense surgery, as each tiny blood vessel must be reattached to ensure proper blood supply to the tissue.

Types of Free Flaps:

DIEP Flap – stands for Deep Inferior Epigastric Perforator. This is the named vessel for which the tissue to be transferred is based. The deep inferior epigastric vessels arise from the external iliac vessels and course beneath the rectus abdominis on each side. These vessels send off branches to the muscle as well as through the muscle and the overlying fat. These perforating vessels are those which are identified, preserved, and transferred with the overlying tummy fat to reconstruct the breast. You are a candidate for this procedure if the amount of fat you have on your lower abdomen is sufficient to reconstruct one or both breasts to the desired volume. Unlike the TRAM flap, prior abdominal operations does not exclude the DIEP flap from use. A prior tummy-tuck does exclude the DIEP from being used. Most plastic surgeons don't perform perforator breast reconstruction due to it's complexity. It is technically very difficult and time consuming and usually necessitates a microsurgeon. The DIEP is a "muscle preserving" procedure and doesn't sacrifice the rectus abdominis muscle.

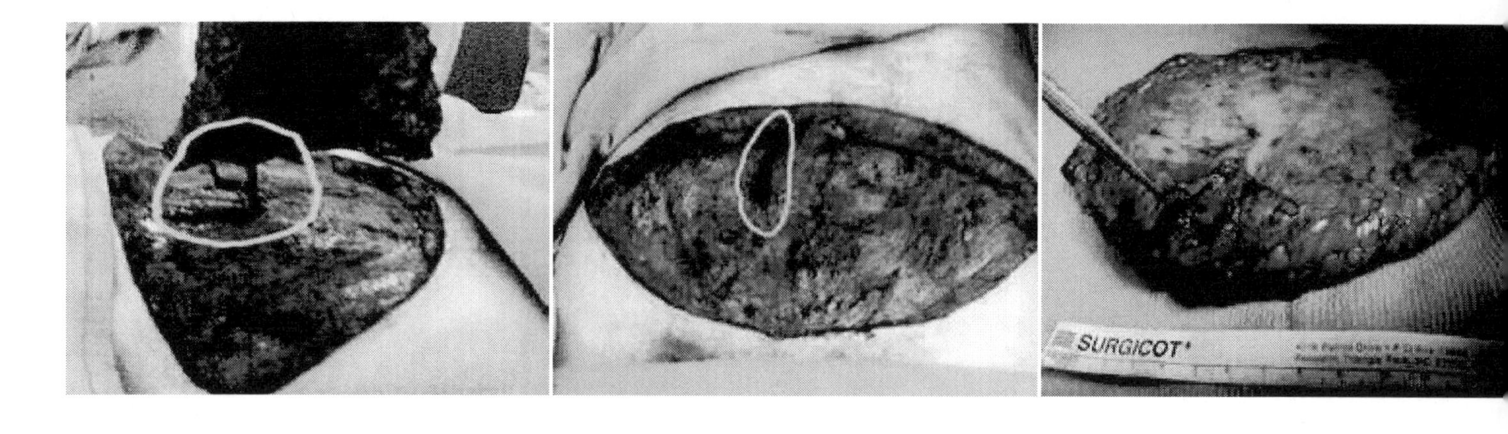

Surgery – Free Flap / attached Surgery – Free Flap / removed Surgery – Free Flap

GAP Flap – stands for Gluteal Artery Perforator. This may be described as S-GAP or I-GAP. The prefixes define superior or inferior branches of the gluteal artery. As with the DIEP, the gluteal artery perforator arises from a branch of the gluteal artery, courses through the muscles, and delivers blood to the overlying buttock fat. This procedure allows for use of buttock fat to reconstruct the breast when abdominal fat is inadequate. Similar to the DIEP it is a "muscle preserving" procedure and doesn't sacrifice the buttock muscle.

I-GAP:

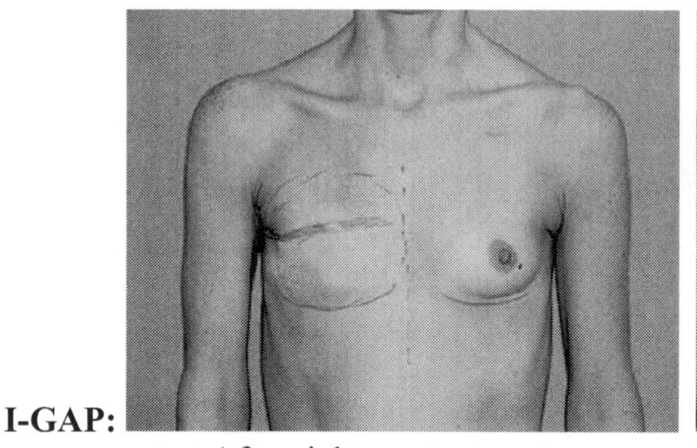

After right mastectomy

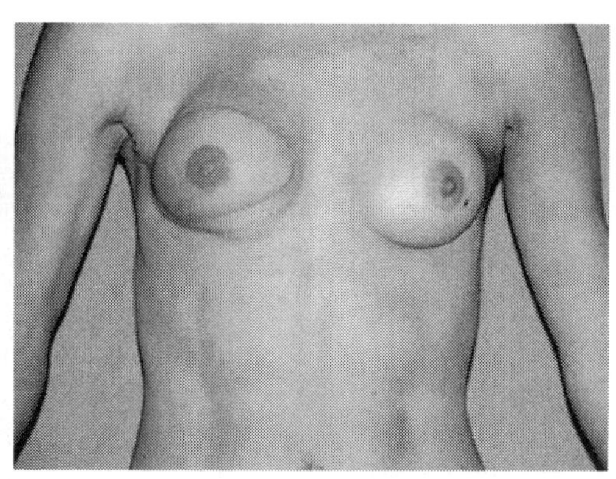

Five months later

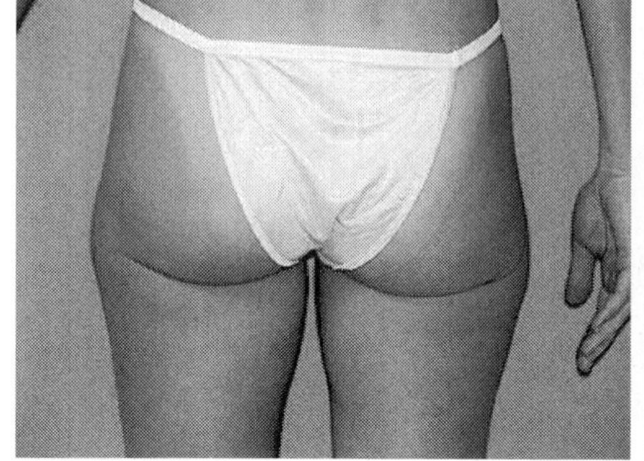

Before right buttock harvest

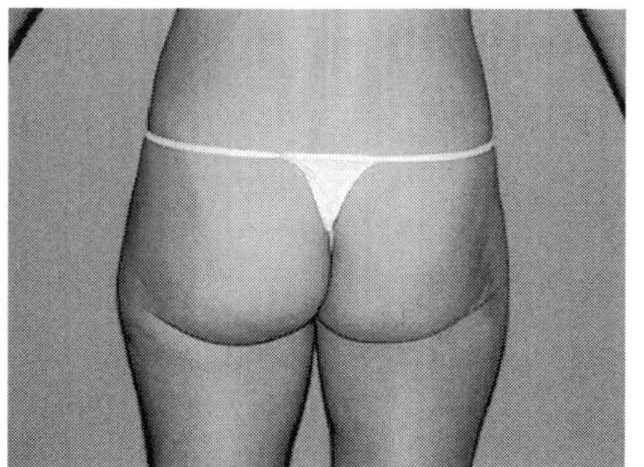

Five months later

S-GAP:

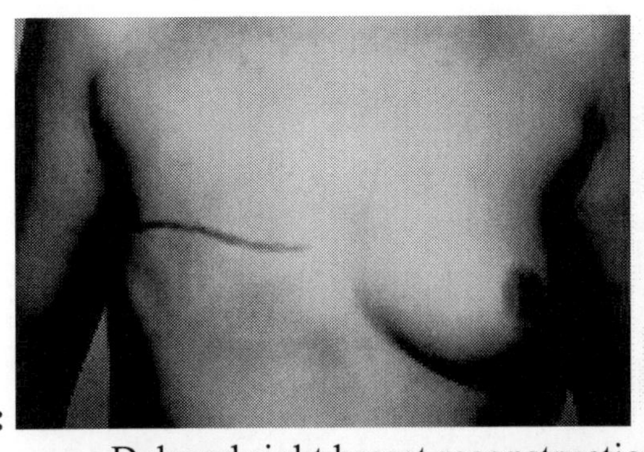

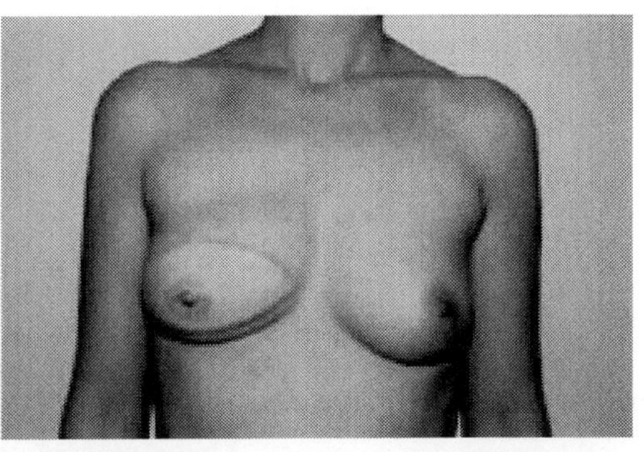

Delayed right breast reconstruction with S-GAP

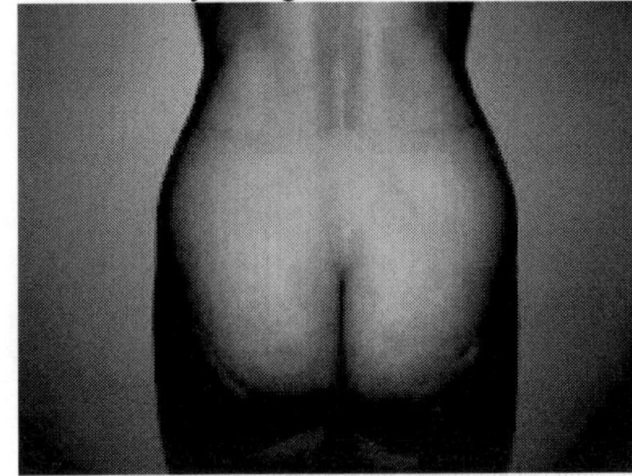

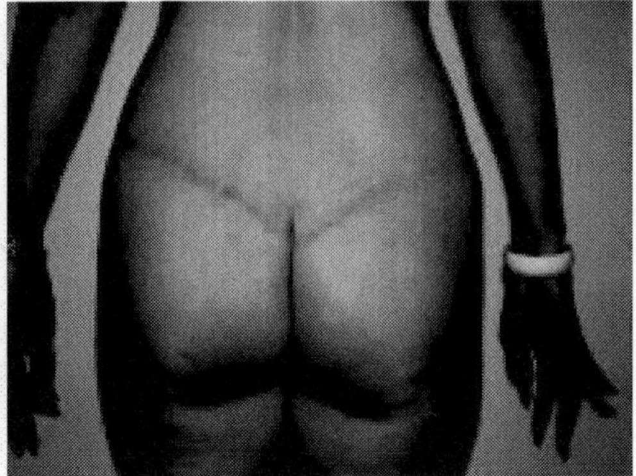

Areola reconstruction

Skin grafts: A 2-dimensional structure, the areola usually does not pose the same degree of difficulty in reconstruction as the nipple. Time-honored methods of reconstruction included skin grafts obtained from the contralateral nipple-areola complex or the medial thigh-vulva area. These grafts were intended to provide pigmented tissue that closely resembled the native areola. Currently, with great improvements in tattoo technique and seemingly endless variety of pigments available, it seems unjustifiable to expose a patient to the risks of possible donor site complications to obtain the same or less satisfactory result than one obtainable with tattoo.

Tattoo: Tattooing of the reconstructed breast in the area of the expected nipple areola complex is simple, easy to perform, readily correctable, and requires no significant patient participation or convalescence. Reconstruction of the areola by tattooing can be staged and often requires some touch-up, but it allows a great degree of control over size, definition of the contour, and color match with the contralateral side; however, results are usually less than satisfactory when used in the absence of nipple reconstruction.

Tattooing is strictly an optical effect, with no structural support and no way to reproduce the projection and texture typical of the natural nipple-areola complex. Thus, tattooing is best employed as a stage of the combined approach toward nipple reconstruction, whether performed prior to or after elevation of the nipple structure.

Many tattoo pigments are available, allowing precise color matching with the contralateral areola. Choose a shade just darker than the desired color to allow for some fading, which necessarily occurs in the weeks post injection. One or more applications may be necessary to obtain the desired result. Commercial surgical tattoo equipment usually is adequate to treat thick skin, such as the lower abdominal and back skin transferred with myocutaneous reconstruction flaps, and the injection technique is acquired easily with minimal practice. In some cases it may be advisable to tattoo the skin prior to elevation of the nipple structure, to facilitate later touch-up and obtain a more natural color distribution.

Nipple reconstruction

The most challenging aspect of nipple reconstruction is the creation of a 3-dimensional projecting structure with texture, dimensions, and contour similar to the contralateral nipple. Moreover, the reconstruction has to last. Nipple reconstruction enhances the realism of breast reconstruction, and the more projecting and 3-dimensional the structure, the more lifelike the reconstruction. Various options have included banking, nipple sharing, grafting, and local flaps.

Nipple reconstruction with local flaps is achieved with various techniques, each with its own proponents and benefits. These include the skate flap, bell flap, double opposing tab flap, star flap, top-hat flap, twin flap, propeller flap, S flap, rolled dermal-fat flap, and autologous cartilage.

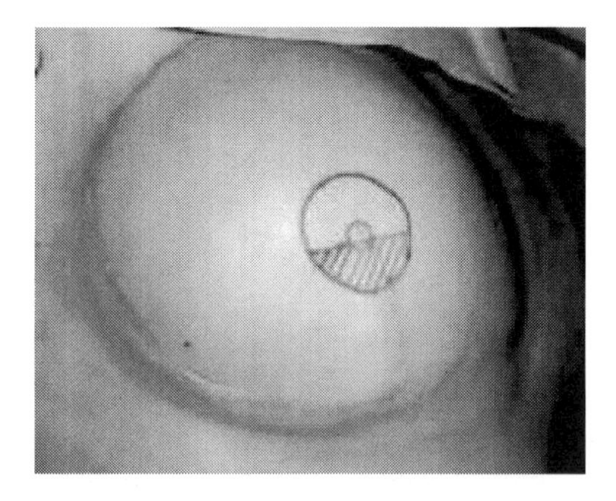

Figure 1 - This drawing represents a typical method of performing a "skate flap" nipple reconstruction.

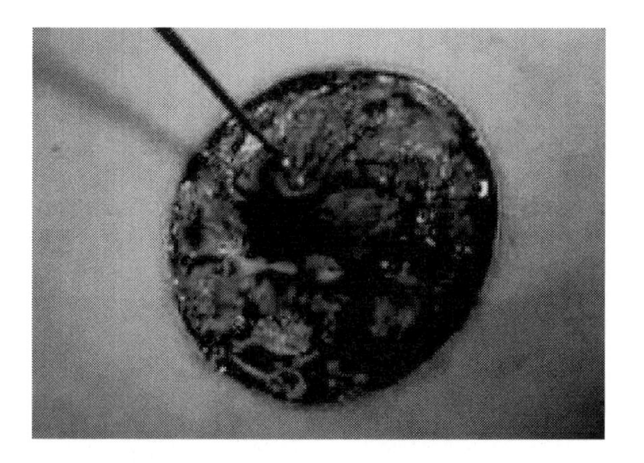

Figure 2 - This photograph shows that the "hatched" area in the previous photograph has been excised.

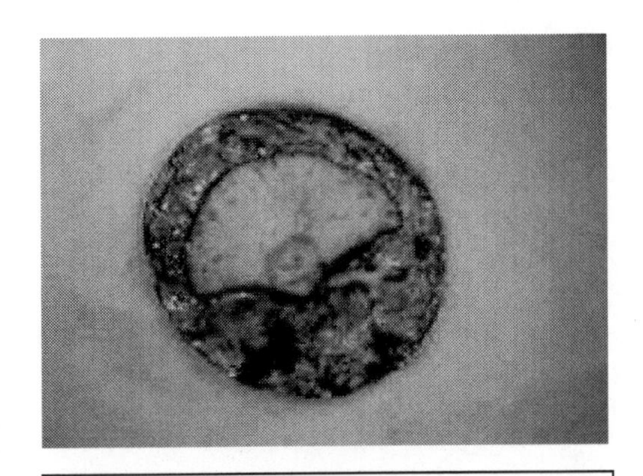

Figure 3 - The remaining skin is then folded onto itself to create a nipple.

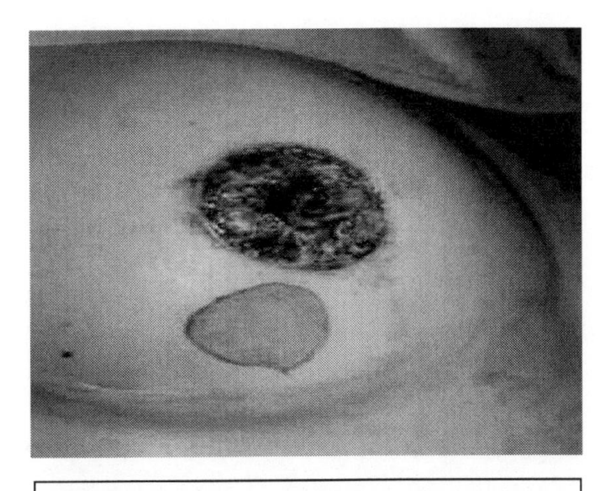

Figure 4 - This photograph shows the newly formed nipple and a graft that will be used for the areola reconstruction

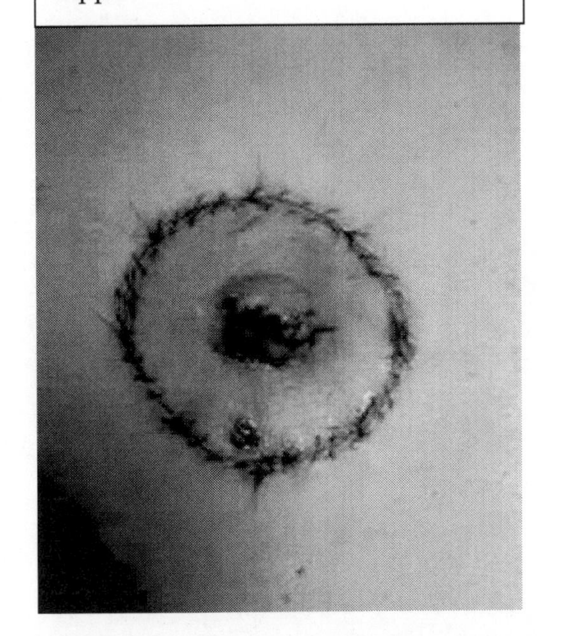

Figure 5 - This is the final result immediately postoperatively.

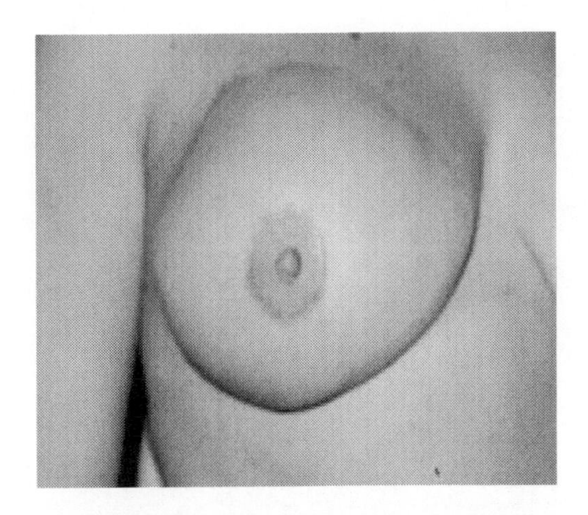

Figure 6 - This is the final result later postoperatively (from the side).

Cancer Treatment

Our goal as health and fitness professionals is to create an exercise recovery program that will cater to the patient/survivors' individual needs both during and after cancer treatment. The initial assessment is critical in determining the best protocol for each individual. It is also important to periodically reassess your patient/client to determine if any changes need to be made in their program.

Many considerations go into customizing a program for the cancer patient/survivor; stage of cancer, type of treatment, time since last treatment, blood profile, medications, environmental factors, and past fitness level to name a few. It is important to understand the following physical changes that may occur during and after cancer treatment:

- Because of the muscle wasting that occurs during cancer treatment, it is essential to keep the exercise at a low to moderate intensity in order to stimulate protein synthesis so that protein will be available for tissue repair. During exercise protein synthesis (the process in which cells build proteins) is triggered for energy use and tissue repair. Cancer patients should exercise at moderate intensity in order to maximize the use of body fat (lipolysis) and minimize the production of lactic acid.

- Both radiation and chemotherapy compromise the immune system. Both lymphocytes and T cells, which are prominent in the immune system, are extremely vulnerable to damage both during and after therapy. The extent of the damage will depend on the type, dose, and localization of the treatment. Patients will be much more susceptible to common bacteria and infection and extraordinary care should be taken to keep the environment as sterile and free of germs as possible. If you have a cold or flu you should ***not*** work with cancer patients until you are better. In addition, patients with low platelets may suffer excessive bruising and bleeding and should be careful not to get any cuts or scratches and try and avoid bumping into things.

- Pulmonary toxicity may be acute or chronic. The effects may be seen within days, months, or even years following radiation and chemotherapy. Long-term treatment can cause the formation of scar tissue and abnormal development of pulmonary tissue. Patients may experience coughing, a low grade fever, low exercise tolerance, restlessness, and severe fatigue. Patients will also suffer with frequent colds and shortness of breath.

- Cancer treatments can also disrupt the normal cell reproduction cycle, leading to acute changes in the intestinal mucosa. Patients may experience thickening or narrowing of bowel segments, ulceration, diarrhea, decrease in the level of digestive enzymes (this may lead to protein deficiency and impaired ability to absorb fat, carbohydrates, vitamins, and electrolytes). Patients may present as extremely thin and weak due to malnutrition.

- Radiation to the chest can lead to pericarditis (inflammation of the membranous sac enclosing the heart). While some patients will recover from pericarditis in 2-5 months following treatment, others require surgery to prevent restrictive inflammation surrounding the heart.

- High doses of radiation can have a detrimental effect on thyroid tissue, leading to serious thyroid malfunction. Radiation can also have an effect on the hypothalamus and pituitary glands, causing abnormalities in the release of HGH.

- Certain types of chemotherapy drugs can directly damage the heart. There is a correlation between increased doses of chemotherapy and a gradual increase in heart damage. Cardiomyopathy typically appears after about four weeks of treatment and has a higher fatality rate than does radiation-induced cardiomyopathy. Signs of chemotherapy-induced cardiomyopathy include:

    - Sinus tachycardia – a heart rate greater than 100 beats per minute in an average adult - originates from the sinoatrial node (the heart's electrical pacemaker)

    - Premature atrial contractions – a type of irregular heartbeat which starts in the upper two chambers of the heart. They are not as serious as premature ventricular contractions. An individual with this condition may report feeling that his or her heart "stops" after a symptom. They are also called heart palpitations.

    - Premature ventricular contractions – a type of irregular heartbeat which starts in the lower chambers of the heart.

    - Supraventricular arrythmias – early extra beats that originate in the upper chamber of the heart

- Chemotherapy drugs such as cisplatin, methotrexate, and mitomycin, can lead to an abnormal amount of uric acid in the blood (uricemia). Side effects may include kidney and bladder problems, hemolytic anemia, edema (particularly in the upper and lower limbs), and decreased levels of magnesium, potassium, calcium, and sodium.

In addition, methotrexate is hepatoxic (causes liver damage). Ongoing changes to the liver can lead to cirrhosis (disease of the liver) and cholestasis (loss of bile excretion). Fortunately, liver disease is usually temporary and controllable with other therapies. Patients present with symptoms such as rapid weight gain, anorexia, fatigue, increase in abdominal growth, and pain.

Radiation Therapy

Types:

External beam - a machine sends high-energy rays directly to the tumor, or if given after surgery has been performed, at the tumor bed, to wipe out any cells left behind. The objective is to destroy the tumor cells with minimal damage to the normal, healthy tissue and organs. On average, patients receive treatment Mon-Fri for 6 weeks. ***Intraoperative radiation*** is external beam radiation given during an operation. It is an experimental treatment being used for cancers of the breast, colon, rectum, stomach, brain, pancreas, and gynecologic organs. Radiation therapy kills cancerous cells while they are undergoing cell division. Normal tissue that functions without constant cell reproduction such muscle, neural, reproductive, and lung tissue, skin, gastrointestinal mucosa, bone marrow, and exocrine glands, is more resistant to damage from radiation. On the other hand, the normal vascular and connective tissues that support muscles and nerves does, in fact, reproduce frequently, making them more susceptible to the effects of radiation.

It is important that your clients do not apply lotions or jellies that contain petroleum jelly as they may interfere with treatment or healing. Patients should consult their doctor or nurse for product recommendations. It is also important not to apply anything hot or cold to the affected skin without first consulting the radiation oncologist. Heat or cold may further irritate the already sensitive skin. If your client goes in a swimming pool or salt water, they must rinse their skin well with fresh water. Skin should not be exposed to direct sunlight for at least one month after treatment.

Potential side effects:

All sites:

- Mild to moderate pink color, or redness, and with itching, burning, soreness, and possible peeling.

- The skin might have a more dramatic reaction to radiation if patients' complexion is fair and susceptible to sunburn or if they've had recent chemotherapy.

- Skin may also become dry, sore, and more sensitive to the touch.
- The skin can start to peel in a dry way- like an old sunburn, or in a wet way, like a blister.
- If the blister opens, the exposed raw area can be quite painful and weepy. The skin reaction can become more serious if the exposed area is not treated, and infection develops.
- The deep redness and the sensitivity should go away over the first weeks after treatment. The skin will take somewhat longer to completely return to its natural color.
- Patients may find that the treated area has a tanned or slightly pinkish look to it for up to six months after their last session of radiation.
- If patients are African-American, their skin may be more tan and less red. The skin can become very dark, and it may take longer for changes to go away.
- Hair loss in area being treated
- Sluggish bowels
- Mental and physical fatigue
- Lymphedema in the area that was irradiated

Head, neck, upper chest, mouth and throat:

- Thick saliva
- Dry mouth
- Sore throat, red tongue, white spots in mouth, sore mouth, unable to wear dentures, lumplike feeling when swallowing
- Loss of taste or change in taste
- Problems with teeth
- Earaches
- Drooping or swelling skin under chin
- Loss of hair

Breast:

- Dry, tender, moist, or itchy skin in armpit, under breast, or breast area

Upper abdomen:

- Nausea, vomiting, feeling of fullness

Lower abdomen:

- Diarrhea, feeling sick to your stomach, cramps, rectal burning with bowel movement, and inflamed bladder

Pelvis:

- Infertility or sterility
- Vaginal dryness
- Painful intercourse
- Impotence
- Low sexual desire

Brachytherapy (internal) – is also called seed implantation or interstitial radiation and it entails surgically implanting a radioactive substance into a body cavity or directly into the tumor. There are two types:

- **Permanent (low dose rate or LDR)** – seeds of radioactive material are placed inside thin needles which are then inserted into the skin. This is usually done under general anesthesia, or a spinal nerve block. The seeds are left in place and the needles are removed. They give off low dose radiation for weeks, or even months. Because the radiation from the seeds doesn't ravel far, they can put out a very large amount of radiation to a very small area. This minimizes damage to the surrounding healthy tissue. After the radioactive material is used up, the seeds are just left in place. Because they are so small they cause little to no discomfort.

- **Temporary (high dose rate or HDR)** – hollow needles are inserted into the skin and catheters are placed in the needles. The needles are removed and the catheters stay in place. There may be some pain in the area of placement for about a week. The radioactive substance is placed in the catheters and delivered for 5-15 minutes. Typically three treatments are given over a couple of days and the radioactive substance is removed every time. After the last treatment the catheters are removed.

These treatments are usually combined with external beam radiation, however the external beam radiation is given at a lower dose than if it were given by itself. Once a patient has received \ radioactive seeds they will give off a small amount of radiation for several weeks. Although the radiation doesn't travel far, the patient should stay away from pregnant women and small children during that time.

Three-dimensional conformal radiation therapy – uses special computers to map out the exact location of the cancer. The patient is fitted with a plastic mold that resembles a full body cast. The mold will keep them in position so that the radiation can be administered more accurately. Radiation beams coming from several directions are aimed at the target, minimizing the risk of damaging the healthy surrounding tissue. This method seems to be as effective as external beam radiation therapy with less damage to the healthy tissue.

Conformal proton beam radiation therapy – uses a similar approach to the 3-D conformal radiation. Instead of using x-rays it focuses proton beams on the cancerous cells. Unlike x-rays which release energy both before and after they hit the targeted area, protons cause little damage to tissue because they pass through the tissue and don't release their energy until they have traveled some distance past the target. Proton beam therapy is still in somewhat of a trial phase and not readily available. There are only a handful of these very expensive machines in the U.S. and they may not be covered by insurance at this time.

Intensity-modulated radiation therapy (IMRT) – may be used to decrease toxicity to normal tissue. IMRT is a type of 3-D radiation therapy that targets tumors with greater precision than conventional radiation therapy. Using highly sophisticated computer software and 3-D images from CT scans, the radiation oncologist can develop an individualized treatment plan that delivers high doses of radiation to cancerous tissue while sparing surrounding organs and reducing the risk of injury to healthy tissues.

Stereotactic radiosurgery - by surgically implanting radioactive pellets into the tumor site, a lethal dose of radiation is delivered, minimally affecting nearby brain tissue. This procedure is done by creating a small hole in the skull in which the *dionuclides* (pellets) are placed and subsequently removed after 6 or 7 days. Using three-dimensional guidance, the machine sends multiple beams of high-dose radiation to the tumor, blasting it from different angles. Next, the patient is repositioned on the table with a five-hundred-pound helmet containing 201 holes – one for each beam of radiation delivered by the *gamma knife.* This procedure seems to be most effective for small primary and metastatic brain tumors that measure less than 4 cm in diameter, and for those tumors that are considered inoperable.

Potential side effect of stereotactic radiosurgery:

- Nausea
- Neck stiffness
- Pain at the pin sites
- Radiation injury to brain tissue surrounding the target that may cause swelling 3-12 months after the procedure. In most cases it is temporary and resolves itself, but some patients may need steroid medications to control persistent swelling.

Radioactive Iodine Therapy - is given in either liquid or pill form. Radioactive iodine is given to destroy any normal thyroid tissue. This allows the physician to maintain surveillance with blood tests to check a blood marker called thyroglobulin. As the radioactive iodine travels through the body, it is able to find and destroy any thyroid cells that were not removed by surgery or those thyroid cancer cells that have spread beyond the thyroid. Usually only one or two treatments with radioactive iodine are necessary. In preparation for radioactive iodine treatment, most patients must stop taking thyroid replacement pills for an appropriate period of time before beginning radioactive iodine treatment to help their bodies produce more natural TSH.

Potential side effects of radioiodine therapy:

- Temporary loss of taste
- Pain swelling and tenderness of the salivary glands
- Temporary cessation of menstruation in older women
- Decreased fertility in men
- Headache, nausea, and vomiting at high doses
    - If cancer cells have metastasized to the bladder:
        - Frequent urination
        - Discomfort while urinating
        - Bloody urine

– If cancer cells have metastasized to the stomach:
    - Abdominal pain
    - Nausea, vomiting
    - Appetite loss
– If cancer cells have metastasized to the central nervous system:
    - Swelling of the brain (cerebral edema)
    - Compression of the spinal cord

Chemotherapy

Chemotherapy – is the administration of anti-tumor drugs that destroy cancerous tumor cells. It can be given as the primary therapy for advanced cancers, or, as an adjuvant therapy with other localized treatments such as radiation. As with radiation, the goal is to destroy the cancerous cells and minimize the damage, and limit the disruption of the normal cells. However, due to the similarity between normal and cancerous cells, it becomes very difficult to destroy the cancerous cells without affecting normal cells. Chemotherapy attempts to target rapidly dividing cells. Because cancerous cells are dividing more rapidly than normal cells, chemotherapy destroys a higher percentage of cancerous cells than normal cells. There are several categories of chemotherapy drugs:

- ***Alkylating agents*** – these agents bond with the DNA in the cancerous cells to prevent reproduction. They attack all cells in a tumor, whether they are dividing and reproducing or not.
- ***Antimeabolites*** – these agents attack the cancerous cells during cell division and eventually cause the cancerous cells to starve to death
- ***Antitumor antibiotics*** – disrupts the role of cancer cell synthesis within the cell by inserting into strands of DNA and either breaking the chromosomes, or inhibiting the synthesis of RNA
- ***Alkaloids*** – interrupt the formation of chromosomes, therefore preventing cell duplication

Administration:

- **Drip** – chemotherapy is commonly given as a drip into a vein in the hand or arm. It may be administered weekly or every couple of weeks, depending on the protocol. An implanted port may be implanted under the skin surface, most commonly in the chest area, by a surgeon. It can remain in place and be used for many years. It does not need care at the external site.

- **Intrathecal** - is used for high-grade lymphatic cancers, or cancers of the brain and spinal cord. Because of the blood brain barrier, chemotherapeutic drugs are unable to reach the brain and spinal cord in adequate concentrations due to the network of blood vessel walls that protect the central nervous system from viruses and toxins in the bloodstream. The chemotherapy is injected directly into the cerebrospinal fluid that surrounds the brain and spinal cord, bypassing the *blood brain barrier.*

- **Immunochemotherapy** - combines chemotherapy and the immune response modifier – interferon, for advanced low-grade lymphomas.

    Potential side effects:

    - Nausea, vomiting, constipation, and diarrhea
    - Hair loss
    - Fatigue / anemia
    - Pain
    - Immunocompromization
    - Infections
    - Bleed and bruise easily
    - Mouth sores
    - Taste and smell changes
    - Appetite changes
    - Skin and nail changes
    - Fluid retention
    - Changes in the smell and color of urine

- Frequent need to urinate
- Instant menopause - menopausal symptoms
- "Chemo-brain"
- Diabetes and osteoporosis
- Damage to the heart and lungs
- Shingles (herpes zoster)
- Peripheral neuropathy of the feet and hands
- Infertility

The following may signal low blood cell counts and should be reported immediately to the doctor:

- Unexpected bruising
- Small red spots under the skin
- Pink or reddish urine, or black or bloody bowel movements
- Bleeding from gums or nose

The following may be signs of an infection and should be reported immediately to the doctor:

- Fever
- Chills, cough, sore throat
- Sweating
- Loose bowels or burning feeling with urination
- Unusual vaginal discharge or itching
- redness or swelling, especially around a sore, pimple, or boil

Cryosurgery

Uses cold-tipped probes filled with liquid nitrogen or carbon dioxide to destroy cancerous tissue. A patient who has a small localized tumor that did not respond to radiation therapy and who could not withstand other treatments, would be eligible for this procedure because it is minimally invasive. Because it is not as invasive as other procedures there is less pain, bleeding, and other complications.

Potential side effects of cryosurgery:

- When used to treat the prostate:
    - It may obstruct urine flow or cause incontinence (these side-effects are usually temporary)
    - Impotence
    - Injury to the rectum
- When used to treat the liver there may be damage to the bile ducts and major blood vessels leading to hemorrhage or infection
- Cryosurgery for cervical intraepithelial neoplasia has not been shown to affect a woman's fertility, but it can cause cramping, pain, or bleeding.
- When used to treat skin cancer (including Kaposi's sarcoma):
    - scarring and swelling can occur
    - If nerves are damaged, loss of sensation may occur
    - Rarely, it may cause a loss of pigmentation and loss of hair in the treated area.
- When used to treat tumors of the bone:
    - It may lead to the destruction of nearby bone tissue and result in fractures
    - These effects may not be seen for some time after the initial treatment and can often be delayed with other treatments.

Photodynamic therapy

It entails injecting a photosensitizing agent into the circulation. Ironically it does not treat cancer systemically, but remains local. Although the drug enters all of the body's cells, it clears from non cancerous cells rapidly. Forty-eight hours after the injection, a red beam of argon laser light is sent through a fiber optic scope placed against the tumor, setting off a chemical reaction that destroys the cancer cells. This treatment may also be used palliatively for the treatment of tumors that are obstructing the esophagus and interfering with swallowing.

Potential side effects:

- Photodynamic therapy makes the skin and eyes sensitive to light for 6 weeks or more after treatment.
  - Patients are advised to avoid direct sunlight and bright indoor light for at least 6 weeks.
  - If patients must go outdoors, they need to wear protective clothing, including sunglasses.
- Other temporary side effects of PDT are related to the treatment of specific areas and can include coughing, trouble swallowing, abdominal pain, and painful breathing or shortness of breath.

Testosterone ablation therapy (medical castration)

Hormone therapy is effective because prostate cancer is dependent on testosterone in order to grow. Removing testosterone from the body, either surgically or by blocking testosterone's effect on prostate cancer medically can be an effective way to block cancer growth. By itself it cannot cure patients of prostate cancer and eventually the prostate cancer finds a way to grow without testosterone present. Hormone therapy is also called androgen deprivation or androgen ablation therapy and is usually used when prostate cancer is advanced, either involving the lymph nodes or has spread to other parts of the body. In some instances hormone therapy is used after or before potentially curative therapy in order to shrink the tumor and improve the cure rate, and is then called "adjuvant" or "neoadjuvant" hormonal therapy.

Orchiectomy- One means of removing testosterone is through surgical removal of the testicles, which make the majority of the body's testosterone. This is known as an orchiectomy or surgical castration. Advantages of an orchiectomy are that it is a simple procedure, often making it possible for patients to go home the same day. The procedure is not reversible.

Potential side effects:

- Hot flashes or flushing
- Infertility
- Decreased sexual desire
- Potential impotence
- Breast tenderness or growth
- Osteoporosis and reduction of muscle tone over prolonged periods.

Types of drugs

- **LHRH agonists** – Medical suppression of testosterone's action can take several forms. Luteinizing hormone releasing hormone (LHRH) analogs – these medications block the body's ability to stimulate testosterone synthesis. They are as effective as orchiectomy but require injections on a monthly to every three month schedule. Occasionally these LHRH medications are combined with the second class of medications to improve suppression of testosterone. Side effects are very similar to those associated with orchiectomy. LHRH analogs may also cause a worsening of cancer symptoms, such as bone pain, for 1-2 weeks before the testosterone level begins to fall.

  Goserelin (for men) (gos-er-el-in), *also Known As: Zoladex®, goserelin acetate implant.* It is given subcutaneously (by injection under the skin) in the lower abdomen.

  Early side effects (beginning within one week) include:

  • Pain, redness, or bruising where you got the injection.

  • A temporary increase in bone pain.

  • Difficulty passing urine.

  Late side effects (beginning after one week) include:

  • Hot flashes.

  • Sweats.

  • An initial and short-term worsening of urinary symptoms.

  • Weight gain.

• Swelling and tenderness of the breasts.

• Impotence (decreased erection).

• A decrease in libido (sexual desire).

• Shrinkage of the testicles.

• Emotional changes.

• Fatigue.

• Loss of bone mass.

• Thinning or loss of body and facial (but not scalp) hair.

Have patient/client call their doctor or nurse if they have:

• Worse symptoms.

• A fast heart rate.

• Chest pain.

• Trouble breathing.

• A fever of 101° F (38.3° C) or above.

• Painful urination or trouble urinating.

- **Diethylstilbestrol (DES)** – estrogens suppress the body's ability to produce testosterone, similar to the LHRH analogs and have some effects on prostate cancer independent of their effects on testosterone. Although not widely available, the estrogen diethylstilbestrol (DES) is still used in the treatment of advanced prostate cancer. DES can also cause blockage of veins or arteries, potentially causing severe problems for patients with a history of heart problems or deep venous thrombosis (clots in the legs). Ketoconazole may be used to try to suppress testosterone or other hormone production and must be given with a corticosteroid if administered in high doses.

Potential side effects include:

• Thrombosis (blood clots).

• Fluid retention.

• Fatigue.

• Nausea and vomiting

• Weight gain.

• Swelling and tenderness of the breasts.

• Impotence (decreased erection).

• A decrease in libido (sexual desire).

• Headaches.

• Skin rashes/discoloration.

• Loss of bone mass.

• Thinning hair.

- **Antiandrogen medication** –these medications bind to the testosterone/androgen receptors in the prostate cancer and block their ability to interact with testosterone. These are usually taken orally as pills. In addition they may cause nausea, diarrhea, skin rash or inflammation of the liver.

Flutamide (flew-tuh-mide), a*lso known as: Drogenil®*, it is given orally (by mouth)

Potential side effects:

• Hot flashes.

• Nausea/diarrhea.

• Loss of appetite.

• Swelling and tenderness of the breasts.

• Impotence (decreased erection).

• A decrease in libido (sexual desire).

Have patient/client call their doctor or nurse if they have:

• Worse symptoms.

• Rash.

• Drowsiness/weakness/unusual fatigue.

• Swelling of the body (edema).

• Mood changes.

• Easy bleeding or bruising.

• Persistent sore throat.

• Fever

• Bluish skin.

• Yellowing eyes or skin.

• Dark urine.

• Stomach pain.

• Sunburn-like effect.

Bicalutamide (bye-kah-lew-tuh-mide), a*lso known as: Casodex®*, it is given orally (by mouth)

Potential side effects:

• Hot flashes.

• Nausea/diarrhea/constipation.

• General body aches.

• Increased urination.

• Cough.

• A decrease in libido (sexual desire).

• Swelling or tenderness of the breasts.

Have patient/client call their doctor or nurse if they have:

• Worse symptoms.

• Trouble breathing.

- Swelling of the hands and feet.
- Depression.
- Unusual weakness.
- Fever/chills
- Yellowing eyes or skin.
- Bloody or dark urine.
- Stomach pain.

Nilutamide (nye-loo-tuh-mide), *also known as: Nilandron®*, it is given orally (by mouth)

Potential side effects:

- Hot flashes.
- Nausea/constipation.
- Loss of appetite.
- A decrease in libido (sexual desire).

Ovarian ablation therapy

Hormonal therapy is not a cure, but can minimize the cancer's spread, often for a period of years. It may also be used palliatively to shrink a tumor that is causing pain. The purpose of hormonal ablation therapy is to stop the ovaries' production of estrogen, which causes the breast tumor cells to grow. This can be done through a *bilateral oophorectomy* (removal of the ovaries), or hormonal therapy such as goserelin or another LHRH agonist. Today, women with a known breast cancer gene abnormality may chooses ovarian ablation to reduce the risk of both breast and ovarian cancer. Breast and ovarian cancer risk falls by about 50% when the ovaries are removed. The risk doesn't go down to zero because ovary-like cells that are normally present in the pelvic area could still form a cancer, even after the ovaries are gone.

Three main approaches are used:

- The ovaries can be surgically removed through small incisions, using an instrument called a laparoscope (bi-lateral salpingo oophorectomy).
- Low-dose radiation therapy can be used to completely shut down the ovaries' production of estrogen.
- Hormonal therapy

Types of hormonal therapy

ERD's - estrogen receptor downregulators :

The drug is approved for treating hormone-receptor-positive metastatic breast cancer in post-menopausal women with cancer that is no longer responding to hormonal therapy such as Tamoxifen. ERDs work by attaching to the hormone receptors on breast cancer cells, blocking them, and causing them to break down and stop working. In addition to binding to and blocking estrogen receptors, ERDs also stop or slow down the growth of breast cancer cells by breaking down the receptors. With fewer hormone receptors available, fewer cells receive the signal telling them to grow, and the overgrowth of cancer cells can be slowed or stopped. Because FASLODEX is administered intramuscularly, it should not be used in patients with certain blood disorders or in patients receiving anticoagulants (blood thinners).

Fulvestrant (ful-VEST-rant), a*lso Known As: Faslodex®,* is given intramuscularly (IM), by injection into the buttock.

Potential early side effects (beginning within one week) can include:

• Pain at insertion site for 12 to 24 hours after injection.

• Pain or swelling.

• Cramps or bloating.

• Change in bowel habits.

• Headaches.

• Mild nausea.

• Generalized body aches.

Potential late side effects (beginning after one week) can include:

• Hot flashes.

• Fatigue or lack of energy.

• Sore throat or dry mouth.

SERMs - selective estrogen receptor modulators:

Block the actions of estrogen in breast tissues and certain other tissues by "occupying" the estrogen receptors on cells. The SERM fits in the estrogen receptor, but it does NOT send messages to the cell nucleus to grow and divide. SERMs do send estrogen-like signals when they land in receptors' bone cells, liver cells, and elsewhere in the body. This means that SERMs seem to help prevent or slow osteoporosis in post-menopausal women and may help lower cholesterol. This dual effect—blocking estrogen in some places and imitating estrogen in other places—allows SERMs to have multiple beneficial effects in many women with breast cancer.

Raloxifene (ra-LOX-ih-feen), also known as; Evista®, is given orally (by mouth)

Potential side effects:

- Hot flashes.
- Leg cramps.
- Blood clots in the legs, lungs, or eyes.
- Trouble breathing.
- Chest pain.
- Vision changes.

Tamoxifen (ta-MOX-ih-fen), ***also known as: Nolvadex®,*** is given orally (by mouth).

Potential immediate side effects (beginning within 24 hours) can include:

• Mild nausea.

Potential late side effects (beginning after one week) can include:

- Hot flashes.
- An increase in bone or tumor pain.
- Menstrual periods that become irregular or stop.
- Vaginal dryness or discharge.
- A temporary swelling or puffiness of hands and feet.
- Mood changes.
- Blood clots.
- Cataracts.

Aromatase Inhibitors:

Lower the amount of estrogen being produced by the body. This method contrasts with that of SERMs or ERDs, which block estrogen's ability to "turn on" cancer cells. Limiting the amount of estrogen produced means there is less estrogen available to reach cancer cells and make them grow. In post-menopausal women, estrogen is no longer produced by the ovaries, but is converted from androgen, another hormone. Aromatase inhibitors keep androgen from being converted to estrogen. That means less estrogen in the bloodstream and less estrogen reaching estrogen receptors to trigger trouble. Aromatase inhibitors are used primarily for post-menopausal women with metastatic breast cancer. They cannot stop the ovaries of premenopausal women from producing estrogen; for this reason they can only be used in postmenopausal women. In the past, these medications were most commonly used by women who may have already tried other anti-estrogen therapies, such as Tamoxifen, and whose cancer was no longer controlled by those drugs. Now with the results of new studies, many doctors recommend an aromatase inhibitor BEFORE Tamoxifen for post-menopausal women with metastatic disease.

Anastrozole (an-AS-troh-zole), *also known as: Arimidex®*, is given orally (by mouth)

Potential early side effects (beginning within one week) include:

- Headaches.
- Body or joint aches or stiffness.
- Mild nausea.

• Mild stomach cramps or bloating.

Potential late side effects (beginning after one week) include:

• Fatigue.

• Hot flashes.

• Vaginal dryness.

• Decreased bone density.

• Hair thinning.

Letrozole (LET-ro-zole), a*lso known as: Femara®,* is given orally, by mouth.

Potential early side effects can include:

• Headaches.

• Mild nausea.

• Body or joint aches or stiffness.

• Mild stomach cramps or bloating.

Potential late side effects can include:

• Fatigue.

• Vagina dryness.

• Hot flashes.

• Decreased bone density.

Exemestane (eks-e-MESS-tayn), a*lso known as: Aromasin®,* is given orally (by mouth)

Potential early side effects (beginning within one week) can include:

• Headaches.

• Mild nausea.

• Body or joint aches or stiffness.

• Mild stomach cramps or bloating.

Potential late side effects (beginning after one week) can include:

• Fatigue.

• Vaginal dryness.

• Hot flashes

• Decreased bone density.

Leuprolide Acetate for Women (LOO-pro-lyde ASS-e-tate) - ***also known as: Lupron®,*** **is given by injection.**

Leuprolide is used to stop your ovaries from working. It does this by blocking hormone signals from the brain to the ovaries. This results in a drop in your estrogen levels. Lowering estrogen can help treat some medical conditions. It is given **i**ntramuscularly (IM), into muscle.

Potential side effects:

• Hot flashes and sweating

• Pain or swelling at injection site

• Decreased sexual desire

• Vaginal dryness

• Urinary symptoms

• Irregular or no menstrual periods

• Tiredness

• Mood changes

• Decreased bone density

• Headache

• Mild nausea (rare)

Hyperthermia

Hyperthermia heats the tumor (*local hyperthermia*), an organ or limb (*regional hyperthermia*), or the entire body (*whole-body hyperthermia*) to between 40 and 43 degrees Celsius, making them more susceptible to radiation therapy and chemotherapy.

Potential side effects of hyperthermia:

- Discomfort or even significant local pain in about half the patients treated.
- It can also cause blisters, which generally heal rapidly.
- Less commonly, it can cause burns.

Autologous bone-marrow transplantation

Once the cancer is in remission, a solid tumor may benefit from autologous bone-marrow or peripheral–blood stem-cell transplantation as a complement to high-dose chemotherapy. Some patients with leukemia, lymphomas, and multiple myeloma may also be candidates for this type of transplantation if an allogeneic donor can't be found. Prior to the BMT, the patient will receive intense chemotherapy and maybe even radiation. These high dose treatments will destroy the healthy marrow as well as the cancer cells. This pre-treatment phase may last from 2-10 days, depending on the procedures being used. The hospital stay will be approximately one month with an additional 20-25 days for the platelets and white blood cells to reach desirable ranges. There's very little chance of rejection since autologous BMT gives you back your own stem cells and Graft Versus Host Disease (the donor marrow perceives the host as foreign and attacks it) reactions are extremely rare. Because there is no need for immunosuppressant drugs, the immune system rebuilds much more rapidly.

Potential side effects of autologous BMT:

- Graft rejection
- Graft failure-very rarely the new stem cells may not produce enough blood cells and will require another transplant
- Immunocompromization/infection
- Graft Versus Host Disease – can happen up to 6 months after the transplant

Potential side effects of GVHD

- Soon after transplant it would be evident if GVHD has affected the lungs or digestive system
- Severe skin rashes, loss of elasticity, and tightness
- Shortness of breath
- Feeling sick and vomiting and/or diarrhea
- Muscle weakness
- Jaundice

Allogeneic bone-marrow transplantation

Marrow comes from a genetically compatible donor and is the preferred method for leukemia. The higher the donor match, the greater the chances of the graft "taking." Closely matched marrow reduces the chances of the transplanted immune cells attacking the tissues of the recipient (host), a common and potentially fatal complication known as *graft-versus-host disease.* Most patients will stay in the hospital, in a protective isolation room, for a month or two after the transplant has taken place. The catheter will be left in place for three to four months after the transplant. This procedure is not recommended for patients fifty-five and older. This is the most complicated and most expensive of all three procedures.

Potential side effects of allogeneic BMT:

- Graft rejection
- Graft failure very rarely the new stem cells may not produce enough blood cells and will require another transplant
- Graft-versus-host disease (GVHD) – can happen up to 6 months after the transplant

  Potential side effects of GVHD

  - Soon after transplant it would be evident if GVHD has affected the lungs or digestive system
  - Severe skin rashes, loss of elasticity, and tightness
  - Shortness of breath
  - Feeling sick and vomiting and/or diarrhea

- Muscle weakness
- Jaundice
- Bleeding of the nose, mouth, under the skin, or gastrointestinal tract
- Liver damage
- Cataracts in 75% of patients who have single-dose total body radiation
- Immunocompromization

Peripheral Stem Cell transplantation

Stem cell transplant is similar to bone marrow transplant except the cells are collected from stem cells in the bone marrow and then they are frozen. The cells used for transplant can be your own healthy cells (autologous transplant), or they can be collected from a compatible donor (allogeneic transplant). Physicians harvest bone marrow from the pelvic bones or hip in an operating room while the donor is under general anesthesia. The physician inserts a hollow needle into the rear and sometimes the front hipbone, both of which contain a large quantity of bone marrow. The breastbone is another accessible site that is rich in marrow, but this is very rarely used for harvest. The physician often must pierce the bone in several spots to obtain enough marrow for a transplant. The donor will not need stitches but will have some pain and tenderness at the site of the harvest for about a week. The patient then receives a high dose of chemotherapy, which destroys tumor cells but also destroys most or all of the stem cells in the patient's bone marrow. The harvested stem cells are then administered, or transplanted, to help regenerate the patient's blood and immune systems. This procedure is used more frequently than bone marrow transplant because of shortened recovery times and possible decreased risk of infection. Compared to the month long hospital stay following an autologous BMT, there is usually a 10 day hospital stay and an additional 10-16 days for platelet and white blood cell counts to reach desirable ranges.

Potential side effects of peripheral stem cell transplant:

- Graft rejection
- Graft failure – very rarely the new stem cells may not produce enough blood cells and will require another transplant
- Liver or heart problems
- Shortness of breath
- Garlic smell or taste due to the solution used to freeze stem cells

- Graft-versus-host disease (GVHD)

    Potential side effects of GVHD

    - Soon after transplant it would be evident if GVHD has affected the lungs or digestive system
    - Severe skin rashes, loss of elasticity, and tightness
    - Shortness of breath
    - Feeling sick and vomiting and/or diarrhea
    - Muscle weakness
    - Jaundice
    - Bleeding of the nose, mouth, under the skin, or gastrointestinal tract
    - Liver damage
    - Cataracts in 75% of patients who have single-dose total body radiation
    - Immunocompromization

Apherisis

Physicians may collect stem cells from the circulating blood. This procedure, unlike a bone marrow harvest, does not have to be done in an operating room, and the donor does not have to be under general anesthesia. A few days before the procedure, donors are generally given a medication to mobilize or force stem cells from the marrow into the circulating blood. These agents can cause flu-like symptoms and bone pain in the days before and after the procedure. The donor then spends several hours on each of two or three days for the apheresis process. The blood passes through a tube inserted in one of the donor's veins and then through a machine that separates stem cells from other blood cells, which are then returned to the patient. The stem cells collected during the procedure are used immediately or can be frozen and stored.

Potential side effects of apherisis:

- Graft rejection
- Graft failure
- Infection
- Liver or heart problems

- Shortness of breath
- Graft-versus-host disease (GVHD)

  Potential side effects of GVHD:

  - Soon after transplant it would be evident if GVHD has affected the lungs or digestive system
  - Severe skin rashes, loss of elasticity, and tightness
  - Shortness of breath
  - Feeling sick and vomiting and/or diarrhea
  - Muscle weakness
  - Jaundice
  - Bleeding of the nose, mouth, under the skin, or gastrointestinal tract
  - Liver damage
  - Cataracts in 75% of patients who have single-dose total body radiation
  - Immunocompromization

Immunotherapy

Immunotherapy belongs to a group of proteins known as cytokines. They are produced naturally by white blood cells in the body in response to infection, inflammation, or stimulation. They have been used as a treatment for certain viral diseases, including hepatitis B and C. Interferon-alpha has been approved by the FDA and is now commonly used for the treatment of a number of cancers, including:

- Multiple myeloma
- Chronic myelogenous leukemia
- Hairy cell leukemia
- Malignant melanoma

Interleukin 2 (IL-2) is another cytokine with antitumor activity and is frequently used to treat:

- kidney cancer

- melanoma.

***Potential side effects of immunotherapy*:**

- Flu-like symptoms (magnified at greater doses)
- Lack of energy
- Loss of appetite
- Change in taste
- Muscle pain
- Sleeplessness
- Depression
- Tremors/seizures
- Irregular heartbeat
- Loss of memory
- Weight gain
- Anemia
- Angina
- Shortness of breath, cough, lung congestion
- Agitation/anxiety
- Dizziness
- Mild nausea, vomiting, diarrhea, gas, and constipation
- Mild hair loss
- Skin rash/dryness
- Burning at injection site

Arterial Embolization

Arterial embolization attempts to limit the blood supply to the tumor by obstructing a major existing vessel, starving it to death. Chemotherapeutic agents are injected directly into the artery. The chemotherapy is prevented from traveling any further, causing a higher concentration of the drug where it is wanted most.

Potential side effects of arterial embolization:

- Post-embolization pain in the area that was embolized lasting several hours-several days
- Flu-like symptoms

Monoclonal Antibodies

Monoclonal antibodies are artificial antibodies against a particular target (the "antigen") and are produced in the laboratory. As therapy for cancer, monoclonal antibodies can be injected into patients to seek out the cancer cells, potentially leading to disruption of cancer cell activities or to enhancement of the immune response against the cancer.

***Potential side effects of monoclonal antibodies*:**

- Flu-like symptoms (magnified at greater doses)
- Skin Rashes, itching, or hives
- Diarrhea, nausea, and vomiting
- Low blood pressure

More serious side-effects that may cause doctor to stop treatment include (usually happen immediately):

- Very low blood counts
- Heart problems including irregular heart beat, heart failure, and an increased risk of heart attack
- Low levels of magnesium, potassium, or calcium in the blood
- Serious skin rashes that lead to infections
- Bleeding problems

- Immediate reactions to the infusion including shortness of breath, wheezing, hoarseness, fainting, dizziness, blurred vision, nausea, or chest pain/pressure.

Lab Values/Side-Effects and Exercise Programming

Guidelines for determining PT program, exercise prescription and level of activity for patients with cancer. (Adapted from Chapter 18, Exercise for Cancer Patients by JeanneHicks,Theraputic Exercise, 5th Edition, Eds. JV Basmajain and SL Wolf.

Oncology Section APTA Journal Sept 2006

Hemocrit(hct)
Role: % of red blood cells in whole blood. Used to evaluate anemia and abnormal states of hydration. Increase hct may be due to tobacco use, severe lung disease. Low hct may be due to anemia.

Symptoms include:

- Weakness
- Fatigue
- Dyspnea on exertion
- Tachycardia.

Hemoglobin(Hgb)
Role: Reflects the oxygen carrying component of the red blood cell. Indicates severity of anemia. Decrease Hgb is found in anemia. Increase Hgb is found in smokers and people with lung disease.

Platelets
Role: major player in clotting process. As a result of platelet activity, bleeding is stopped and healing begins. High platelets (thrombocytosis) may be caused by iron deficiency, neoplasm, inflammation, and infection. Low platelets (thrombocytopenia) may be causes by liver disease, platelet disorders, viral infections or medication. With low platelet counts there is a high risk for bleeding in the intracranial space. Prolonged bleeding can occur aster surgery or trauma.

White Blood Cells (WBC)
Role: Indicates the functional status of the immune system. They protect the body against infection and aid in the immune system response. High WBC (leukocytosis) may be due to acute/chronic infection or malignancy. Low WBC (leucopenia) is caused by a disease of the immune system or radiation/chemotherapy.

Cell Counts	Normal Values/unit	No exercise when counts are:	Light exercise when counts are:	Resistive exercises when counts are:
Hemocrit	Male 42-52% Female 37-38%	<25% ROM No aerobics No resistance	>25% (Deconditioned) Low impact and intensity aerobics Isometrics Modified resistance	>25% (Fit) Most programs OK
Hgb	Male 14-18g/dL Female 12-16g/dL These values may be lower in the elderly	<8g/dL ROM No aerobics No resistance	8-10g/dL Low impact and intensity aerobics Isometrics Modified resistance	>10g/dL Most programs OK
Platelets	200,000-400,000/mm3	<5,000/ mm3 Range of motion No aerobics No resistance	5,000-50,000/mm3 Low impact and intensity aerobics Isometrics Modified resistance	>50,0000/mm3 Most programs OK
WBC	4,000-10,000/mm3	<500/mm3 Range of motion No aerobics No resistance	>500/mm3 (Deconditioned) Start at low intensity level Low impact and intensity aerobics Isometrics Modified resistance	>500/mm3 (Fit) Start at higher intensity level Most programs OK
Metastatic or bone tumor		>50% cortex involved No weight bearing No exercise	20-50% cortex involved Partial weight bearing ROM No stretching	0-25% cortex involved Full weight bearing Most exercise programs OK

Exercise Intensity

1. Karvonen Method - Takes into account resting HR which is often higher in cancer patients.

Example of a patient that is 60 years of age with a resting HR of 60

220- age(60) =160

Subtract resting HR 160-60=100

50% + resting HR   100 x 50% -60 =110

110 is the target HR.

2. Designate levels:

Subtract the patient's age from 220 and take a percentage as their target HR for exercise.

Level 1: low intensity, for lower functioning patients - 20-40% of their target HR

Level 2: moderate intensity, for patients with ongoing chemo or radiation 40-60% of their target HR.

Level 3: high intensity, for patients who have completed medical treatment 70-80% of their target HR.

Side-effect/complication	Recommendations
Dehydration (vomiting/diarrhea within 24-36 hours)	Rehydrate/electrolytes
Nausea	Exercise at a level they can tolerate
Fatigue	Exercise at a level they can tolerate preferably 20-30 minutes 3-4 times per week.
Bone/joint pain	Avoid high-impact activities, or those with a risk of falling (prevent fracture). Have client/patient follow-up with doctor to eliminate the possibility of metastasis.
Severe weight loss (greater than 35% of their weight)	Low intensity exercise due to loss of muscle mass. Strength training will be beneficial in increasing lean muscle mass.
Fever	No exercising – recommend that client/patient see their doctor to eliminate the possibility of a systemic infection.
Peripheral neuropathy	If it is in the feet, avoid high-impact activities. Also avoid activities that require balance and coordination. If it is in the hands, use machines rather than hand weights.
Dizziness	Avoid activities that require balance and coordination.
"Pitting" edema, or increase in baseline girth measurements	Have client/patient see doctor immediately to evaluate for lymphedema.

ıcting a Postural Analysis

Posture

Poor posture leads to adaptive patterns of movement and balance; thus causing undo stress on the musculoskeletal system, resulting in wear and tear on the joints and increased likelihood of injury. Once the body begins to compensate for poor habits or injury, the body begins to know this as its' new "norm." This leads to patterns of muscle tightness and weakness. For example, for every inch the head moves forward from neutral posture, the weight carried by the lower neck increases by the additional weight of the entire head. (Liebenson C, ed. Rehabilitation of the Spine. Baltimore: Williams and Wilkins, 1996:177.)

According to the late Dr. Vladamir Janda, a Czech neurologist, there are two basic causes of muscle imbalance; structural and functional. Our standards of care in orthopedic medicine are based primarily on a structural approach. They rely on the visualization of structure through modalities such as X-ray, MRI, and/or surgery. When you hear the term "structural lesion," it is referring to damage to physical structures such as bones or ligaments that can be diagnosed through these sophisticated tests and procedures. The structural lesions can be repaired with the proper combination of immobilization, surgery (when necessary), and rehabilitation.

What happens, however, when these diagnostic tests are inconclusive? We find ourselves as both patients and practitioners at a loss. It is frustrating on many different levels as we try to "fix" something that we are being told does not exist. This is where the term "functional lesion" comes in to play. A "functional lesion" is an impairment in the ability of the structure to perform its normal tasks; movement, strength, flexibility, etc... Our dilemma is that this type of lesion is not as easy to treat because the analysis of the problem is not as cut and dry as it is with a structural lesion. As practitioners it is our job to get to the root of the problem and determine the source of origination of the functional lesion. We must analyze the particular dysfunction by looking at the interactions of all the structures and systems that may contribute to the problem. MRI's, X-ray's and surgery are designed to "diagnose" and "cure" the problem, but they overlook an important part of the problem; the source. If we do not identify the source of the problem, it stands to reason, that it can not be cured or rehabilitated

When we use the term "muscle balance," we are referring to the equality of muscle strength between the agonist and antagonist muscles. Without this balance it would be impossible to properly execute movement and function. Muscle balance can also be viewed as the strength and balance between contralateral (right vs. left) muscle groups. If your client is right-side dominant, the strength in their right quadricep may be significantly greater than that of the left quadricep. This can create a chain reaction of problems from the hip to the ankle as the body tries to compensate for and adapt to the imbalance. Sometimes it is an injury itself that can cause a muscle imbalance while other times the muscle imbalance leads to the injury. There are two ways to look at muscle imbalances. The first is biomechanical, caused by repetitive movements and postural deviations that put undo stress on the joints. As our bodies try and adapt to repetitive motions and/or postural imbalances, there will most likely be adaptations in muscles length, strength, and flexibility. As a muscle becomes dominant over its fellow synergists, joint motion will change and can lead to abnormal stress on a particular joint.

The second is neurological. They are not, however, exclusive of one another as both may contribute to the imbalance. Janda believed that many chronic musculoskeletal conditions are the result of faulty motor learning that prevents the motor system from properly reacting or adapting to different changes within the body. The result of this is seen in poor reflexive or mechanical performance. All systems in the human body function automatically except for the motor system. Therefore, Janda believed that chronic musculoskeletal pain and muscle imbalance are a functional pathology (impaired function of the structures rather than damage to the structures) of the CNS. He stated that muscle imbalances often begin after an injury or pathology leads to pain and inflammation. Muscle imbalances may also arise from alterations in proprioceptive input resulting from abnormal joint position or motion. These two conditions can ultimately lead to muscles that become hypertonic (tight), or inhibited (weak). As the body tried to create a new homeostasis, the imbalance becomes "normal" to the CNS and results in a new pattern of movement.

The necessary ingredient for muscle balance is the functional integration of the sensory and motor systems. Sensory information is elicited in motor response through the central and peripheral nervous systems. ***Afferent information*** refers to sensory input to the CNS that plays several roles in creating motor responses (Holm, Inhahl, and Solomonow 2002). These include triggering a reflex response, determining the extent of programmed, voluntary responses, and integrating feedback for automatic motor output to maintain balance for standing and walking.

Postural stability, or balance, is the ability of the body to maintain its center of gravity within its base of support and limits of stability. It is the result of processing and output of information form the PNS and CNS. Balance of agonists and antagonists is necessary in order for ligaments to provide adequate joint stability and equalize pressure distribution. There are two types of stability that we will refer to; ***dynamic and static.*** Dynamic stability is the result of muscular contraction. It is also referred to as ***functional joint stabilization.*** Static stability comes from passive structures such as bony congruity, ligaments, and joint capsules.

"He who treats the site of the pain is often lost," Karel Lewis.

Tensegrity is the inherent stability of structures based on synergy between tension and compression components. In other words, the body rearranges itself in response to changes in load. The principle of a ***"kinetic chain"*** suggests that each part of the body is interconnected. For example, forces in the foot and ankle can affect body parts such as the knees, the hips, and the lower back. Therefore, if something is out of alignment, or functioning improperly, the body will rearrange itself to compensate for the inadequacy.

Postural chains are the position of one joint in relation to the other when the body is in an upright position. We refer most often to the postural chain that occurs throughout the spine. The postural position of the cervical, thoracic, and lumbar spine, is looked at in relation to musculoskeletal pain. We emphasize proper positioning in these areas during exercise to promote safe movement and reduce the risk of injury. Because these regions are connected through the vertebral column, changes in one region can cause a chain reaction leading to changes in another region. Here are some examples:

1) Poor sitting posture encourages a posterior pelvic tilt. This reduces the normal lordotic curve of the spine. It also reverses the normal kyphotic curve of the spine and encourages the forward position of the head that is often associated with poor posture.

2) The rib cage is an important structure to consider because of its direct influence on the position of the thoracolumbar spine. Those who are weak in the diaphragm and deep spinal stabilizers will often elevate the lower rib cage during inspiration as a compensation for breathing. The repetitive and continuous elevation of the ribs leads to posterior rotation of the ribs on the vertebrae at the costovertebral joint (articulations that connect the heads of the ribs with the bodies of the thoracic vertebrae) and to relative anterior rotation of the vertebrae on the ribs. Ideal posture is sacrificed in an attempt to maintain respiratory integrity.

3) An anterior pelvic tilt is associate with tight hip flexors whereas a posterior pelvic tilt is associated with tight hamstrings.

4) Tightness of the upper trapezius from the cervical region influences shoulder joint movement by positioning the scapula upward and downwardly rotated.

5) Foot pronation causes tibial internal rotation, which causes knee valgus (knock-kneed), and hip internal rotation.

Muscular chains are groups of muscles that work together to influence each other through movement patterns.

Myofacial chains rely on the integrity of the muscle fascia as a vital link to multiple muscles acting together for movement. As well as connecting the extremities through the trunk

Neuromuscular chains provide our bodies with critical reflexes for function and protection. These reflexes serve as the basis for all human movement patterns.

Sensorimotor chains are linked neurologically through afferent (input) and efferent (output) systems for function. They are both reflexive and adaptive and provide local and global dynamic stabilization of the joints.

Now that we have a basic understanding of the principles of "chains' and how they work together both in compensation and adaptation, we are going to get even more specific. The CNS is responsible for regulating two groups of muscles within the body; ***Tonic and Phasi***c.

Neurodevelopmental movement patterns can be broken down into these two groups. These terms are not used in the physiological description of muscle fiber (slow twitch vs. Fast twitch). Tonic muscles are dominant and are involved in repetitive or rhythmic activities as well as the upper and lower extremities' withdrawal reflexes. Their main responsibility is flexion. Phasic muscles tend to be predominant in extension and typically serve as postural stabilizers, working against gravity. These systems do not function independently. They work synergistically in posture, gait, and coordinate movement. Therefore, when we refer to muscle balance, we are really referring to the interaction between the tonic and phasic systems. In essence these systems are the "default" motor program of human movement dating back to the beginning of time. The proper balance of the two systems would be seen in someone with "normal" gait and posture. When you watch someone walk you will notice the contralateral movement pattern between the upper and lower body. The right arm and leg will be in flexion at the same time and the same will apply to the left side. With leg extension you will see arm extension on the opposite side. This can be noted in creeping, crawling, swimming, and walking. For you to be skilled at assessing one's posture you must understand which muscles are more prone to tightness vs. which muscles are more prone to weakness.

The **tonic muscles** in the ***upper body*** that will be more prone to **tightness** are:

- Pectoralis major/minor
- Upper trapezius

- Levator scapulae
- Sternocleidomastoid
- Scalenes
- Latissimus dorsi
- Flexors and pronators

The ***phasic muscles*** in the ***upper body*** that will be more prone to ***weakness*** are:

- Middle and lower trapezius
- Rhomboids
- Serratus anterior
- Deep cervical flexors
- Scalenes
- Extensors and supinators

The ***tonic muscles*** in the lower body that will be more prone to ***tightness*** are:

- Quadratus lumborum
- Thoracolumbar paraspinals
- Piriformis
- Iliopsoas
- Rectus femoris
- IT band
- Hamstrings
- Hip adductors
- Soleus
- Tibialis posterior

The *phasic* muscles in the lower body that will be more prone to ***weakness*** are:

- Rectus abdominis
- Gluteus maximus/medius/minimus
- Vastis medialis/lateralis
- Tibialis anterior
- Peroneals

Keep in mind that some muscles may exhibit both characteristics, but will typically be one or the other.

Pain and muscle imbalance

If we injure ourselves it is safe to assume that the particular injury will heal itself in a reasonable amount of time. When someone comes to you with chronic muskuloskeletal pain it may suggest that the problem lies within the muscle, or group of muscles. More often than not it will not stem from the bones, joints, and ligaments. Most muscle pain is the result of a muscle spasm and the resulting ischemia (inadequate circulation to a local area due to blockage of the blood vessels in the area) from the prolonged muscle contraction. This will ultimately lead to fatigue and result in a decreased ability to meet normal postural and movement demands. In the acute phase of pain the muscles may respond by altering
movement patterns to compensate for the injured area. As time passes, the CNS will adapt this altered movement pattern. Our bodies have a protective adaptation to pain in which the flexor response is activated to protect the injured area. Not only will this affect movement patterns, it will also result in decreased range of motion. These altered movement patterns will ultimately lead to altered joint position which will cause more stress the joints.

It is paradoxical that the muscle imbalance may be the source of the pain or it may be the result of the pain.

Vladamir Janda observed three distinct stereotypical patterns of muscle tightness/weakness that cross between the dorsal and ventral sides of the body. The first, and perhaps most common is u***pper-crossed syndrome.*** Tightness in the levator scapulae and upper trapezius on the dorsal side crosses with the tightness of the pectoralis major/minor. Weakness of the deep cervical flexors on the ventral side crosses with weakness of the middle and lower trapezius. This pattern of muscle imbalance creates joint dysfunction that results in forward head, cervical lordosis, elevated and protracted (rounded) shoulders, winged scapulae, and thoracic kyphosis.

This combination wreaks havoc on the glenohumeral joint by decreasing joint stability, which then leads to increased activation by the levator scapulae and upper trapezius, in an effort maintain the integrity of the joint.

Lower-crossed syndrome manifests when the tightness in the thoracolumbar extensors on the dorsal side crosses with the tightness of the iliopsoas and rectus femoris. Weakness of the deep abdominal muscles ventrally crosses with weakness of the gluteus maximus/medius. Look for increased lumbar lordosis, lateral lumbar shift and leg rotation, and knee hyperextension. There are two variations of LCS; anterior tilt and posterior tilt. The client with an anterior pelvic tilt will usually present with slight hip and knee flexion and hyperlordosis of the lumbar spine and hyperkyphosis of the upper lumbar and lower thoracolumbar areas. Those with a posterior pelvic tilt present with locked out knees, hypolordosis (flat back), thoracic hyperkyphosis and head protraction. More than likely they will also have tight hamstrings. And dynamic movement patterns may be affected.

Layer syndrome is the combination of LCS and UCS where patients display impairment with motor skills and have a poorer prognosis because of the longer duration of their impairment. You are more likely to see this in the elderly.

* Muscle imbalance typically begins in the pelvis and moves toward the shoulder and neck area in adults. The reverse applies for children.

Postural Analysis Test Protocol

Purpose: *to determine muscle imbalances; which may cause unnecessary stress on the bones, joints, ligaments, and muscles.*

Equipment: *none*

Procedures: a thorough postural analysis is essential to get a clear understanding of symmetry, contour, and tone of the muscles as they are observed in static posture. The postural analysis is done with the participant swearing minimal clothing, standing erect, naturally, with the arms hanging downward at their sides and bare feet. Have client march in place with their eyes closed for a few seconds, making sure that they are standing naturally. Have them stop marching, hold their position, and open their eyes. Ask them to stand as still as they can so that you can conduct the assessment.

Begin by looking posteriorly at spinal curves; excessiveness, scoliosis, leg-length discrepancy, or other orthopedic deviations. From there you should evaluate the pelvis; as this is where most dysfunctions of the lumbar spine, SI joint, and lower limbs will originate. Not only can the pelvic tilt affect lumbar lordosis, it can also influence the orientation of the head and other parts of the body.

The position of the pelvis should be evaluated by locating the iliac crests and the anterior and posterior iliac spines. Locating the posterior/anterior superior iliac spine:
The ilium is the most superior part of the innominate bone and articulate the pelvis with the spinal column through the sacrum. At the most anterior and posterior aspects of the ilium are bony prominences known as the ***anterior superior iliac spine*** (ASIS) and ***posterior*** *superior iliac spine* (PSIS). The ridge of bone that runs between the ASIS and PSIS, and is a major source of muscle attachments, is known as the crest of the ilium or ***iliac crest.*** Place your hands on your clients' hips and feel for the iliac crests. While you are doing so, follow the crest to its anterior and posterior ends and those will be the ASIS and PSIS.

After identifying the aforementioned structures, there are several things you will want to look for:

- Lateral tilt
- Rotation
- Anterior tilt
- Posterior tilt

Muscle Imbalances and their potential causes:

Postural deviation	Tight musculature	Weak musculature
Forward head	SCM	Deep Cervical Flexors (Longus Coli/Capitus
Elevated shoulder	Upper Trapezius and Levator Scapulae	Lower Trapezius
Shoulder protraction (Round Shoulder Syndrome)	Pectoralis Major/Minor and Latissimus Dorsi	Rhomboids, Middle/Lower Trapezius, Teres Minor, and Infraspinatus
Elbows flex when arms are overhead during squat test	Pectoralis Major	Middle/Lower Trapezius
Arms fall forward when overhead during squat test (increased lordosis)	Latissimus Dorsi and Pectoralis Major	Middle/Lower Trapezius
Winged scapula	Rhomboid	Serratus Anterior
Anterior tilt (increased lumbar lordosis)	Iliopsoas, Rectus Femoris, Erector Spinae, and Latissimus Dorsi	Gluteus Maximus/Medius and Rectus Abdominis
Posterior tilt (increased lumbar flexion-flattened lumbar curve)	External Obliques, Rectus Abdominis, and Hamstrings	Hip Flexors and Quadriceps
Elevated hip	QL (Quadratus Lumborum)	Rectus Abdominis and Obliques
Laterally rotated Patella	TFL (Tensor Fascia Latae), Gluteus Maximus/Medius, and ITB (Iliotibial Band)	Adductors and Sartorius
Medially rotated Patella	Adductors	Gluteus Medius/Maximus
Heels elevate during squat test	Gastrocnemius, Soleus, and Peroneals	Anterior/Posterior Tibialis
Feet externally rotate	Soleus, Biceps Femoris, and Piriformis	Gluteus Medius
Flat foot	Gastrocnemius and Peroneals	Gluteus Medius and Anterior/Posterior Tibialis

Postural alignment checklist

Looking at the subject from the back:

- Head should be erect and not tilted to either side.
- Shoulders should be level and one should not be higher than the other.
- Shoulder blades should not be "winged" and they should be at the same height.
- Curvatures of the spine are minimal (no scoliosis).
- Hips should be at the same level, one should not be higher than the other.
- Legs are vertical.
- Arches of the feet should not be excessively flat or raised, but should appear normal.
- Heels should be equal distance apart.
- Body weight should appear to be evenly distributed on both feet.

Looking at the subject from the front;

- Head should be erect and not tilted to either side.
- Shoulders should be level and one should not be higher than the other.
- Arms and hands should face toward the body. If there is considerable round shoulderedness, it will cause the hands to rotate so that the palms face backward.
- Hips should be at the same level, one should not be higher than the other.
- Legs are vertical and kneecaps should face forward and be centered.
- Toes should be in a straight line.
- Feet should be turned out at about ten degrees.
- Arches of the feet should not be excessively flat or raised, but should appear normal.
- Heels should be equal distance apart.

- Body weight should appear to be evenly distributed over the feet.

Looking at the subject from the side:

- Head should be erect and should not be pulled back or extended forward.
- Shoulders should be level and should not be rounded.
- Curvatures of the spine should be minimal.
- Arms and hands should face towards the body.
- Chest should be lifted and should not appear depressed or protruding.
- Abdominals should be flat.
- Knees should not be overflexed or hyperextended, but should appear straight.
- Toes should be in a straight line.

Overhead squat assessment:

Have client stand with their feet shoulder with apart (making sure both the feet and knees stay in alignment and point forward). Have them raise their arms overhead with their arms fully extended, but not locking the elbows. Have them squat no lower than 90° and then return to starting position. Watch them repeat this pattern of movement several times.

After watching their movement pattern you can derive the following:
If their elbows flex while overhead during the squat test, they probably have tight ***pectoralis major and weak middle/lower trapezius.***

If their arms fall forward while overhead while doing the squat test, they probably have ***tight pectoralis major and latissimus dorsi and weak middle/lower trapezius.***

Hip Abductor Movement Pattern Test

The hip abductors stabilize the pelvis in the frontal plane and help to prevent a strong adductor torque on the knees. If you noted medial or lateral rotation of the knees and/or hip during the postural assessment or the modified Thomas test, you will want to perform this test on your client.

Begin with the client lying on their side with the bottom leg flexed and the top leg in line with the body. Have the client raise their leg toward the ceiling. The normal pattern of hip abduction is about 20° ***without*** any hip flexion or internal or external rotation. Watch the hip to make sure there is no elevation or rotation.

If the TFL is tight the abduction movement will be combined with hip flexion do to the TFLs' dual action as both a hip abductor and flexor. If the QL contracts prior to 20° of hip abduction, it will result in a lateral hip hike. At this point the QL becomes the prime mover in hip abduction rather than just the stabilizer.
Tightness in either the QL or TFL are often associated with tightness of the IT band and atrophy of the gluteal muscles on that side.

Latissimus Dorsi Muscle Length Test

Because of its many attachments, the latissimus dorsi can medially rotate, adduct, and extend the humerus as well as extend the lumbar spine and cause an anterior tilt in the pelvis. Having a short latissimus dorsi can result in excessively rounded shoulders as well as decreased ROM in shoulder flexion.

Begin with the client lying in supine position with the knees bent and feet flat. Stand beside the table by the arm being tested. Passively raise the arm in flexion toward the top of the table. Normal ROM would allow the arm to extend to 180°. If the latissimus is tight, the arm will fall short of 180° and/or the lumbar spine will go into extension (arch).

Evaluation of Balance

Lateral stabilization of the pelvis is required to maintain proper gait. The stability of the pelvis comes from the gluteus medius, gluteus maximus, and TFL. A single-leg stance test for balance can be analyzed to screen for muscle imbalances and risk of injury. Have your client raise one hip to 45° while bending the knee to 90°.

1) Observe the quality of their movement. Is there any shifting while trying to attain the desired stance? Is there any unevenness in their pelvis or shoulders?

2) The client should be able to hold this stance without any compensations for at least 15 seconds.

3) If you notice any of the following while trying to attain stance position, there may be possible dysfunction.

- Excessive shifting of the pelvis
- Inability to hold the stance for 15 seconds
- Elevated contralateral shoulder
- Elevated contralteral hip
- Medial rotation of the femur

If the knee is medially rotating, it may be that the TFL and medial hip rotators are dominant to the inhibited and weaker gluteus medius/minimus, and hip lateral rotators. We can generalize this by saying that the adductors are dominant to the abductors. In this case, chose exercises that focus on strengthening the hip abductors and avoid hip adduction until the medial rotation seems to be corrected. The reverse would apply if the femur was laterally rotated.

At this point you should do the modified Thomas test to confirm your findings and help you to determine the correct balance of stretching vs. strengthening exercises. The modified Thomas test allows you to assess the lower quarter muscles that are especially prone to tightness. They include the iliacus and psoas major, and the rectus femoris and TFL-ITB. If the hip flexors are tight it will limit hip extension and may cause an anterior pelvic tilt. Couples with the tight hip flexors, you will most likely find weak gluteus maximus.

Modified Thomas test

Have your client sit on the very edge of the table or exercise bench with the coccyx touching the table and one foot on the floor **(Figure 1A).** Help them to carefully roll onto their back while helping to support their midthoracic spine and keeping their knees drawing into their chest. Make sure that their spine is in neutral (not arched) and that their pelvis is in posterior rotation to fix the origin of the hip flexors. They should maintain this position during the evaluation.

While maintaining knee flexion, passively lower the clients' leg until you detect resistance in the leg being tested **(Figure 1B).** Once you determine the resting position, look at the thigh from the side to see if it is parallel (180°) to the table or exercise bench, or if it is less than 180°. The greater the hip extension in the leg being tested, the tighter the iliacus and psoas muscles.

Now look at the angle of the knee. The normal angle should be 90°. If the angle is greater than 0°, it is most likely due to tightness in the rectus femoris. With both the hip flexors and the quadriceps, the degree of the angle will reflect the amount of tightness in that area.

Position yourself at the end of the table or exercise bench, facing your client. Is the femur in a straight line with the patella centered? If the tested leg and respective patella are laterally rotating, it suggests that there is tightness in the TFL-ITB. This can be further confirmed by two things; the presence of a lateral IT groove and, when you passively move the tested leg into neutral, an increase in hip flexion

Figure 1A

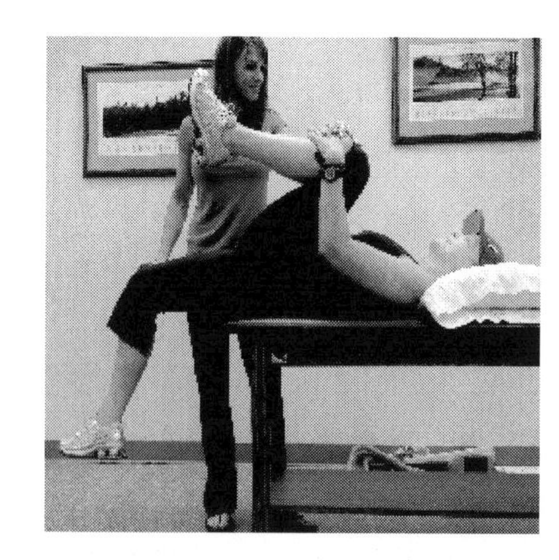

Figure 1B

Using a Goniometer

Range of Motion Assessment Protocol

Purpose*: to determine the range of motion and measure the flexibility of a specific joint*

Equipment: *goniometer*

Range of Motion and Pain Variations

1) Normal mobility with no pain – indicates a normal joint with no lesion

2) Normal mobility with pain elicited – may indicate a minor sprain of a ligament or capsular lesion.

3) Hypomobility with no pain – may indicate an adhesion of a particular tested structure

4) Hypomobility with pain elicited – may indicate an acute sprain of a ligament or capsular lesion. When an injury is severe, hypomobility and pain may also be a sign of muscle spasm caused by "guarding" or by muscle strain opposite the side of the movement.

5) Hypermobility with no pain – suggests a complete tear of structure with no fibers in tact where pain can be elicited. It may also be normal if other joints, in the absence of trauma, are also hypermobile.

6) Hypermobility with pain elicited – indicates a partial tear with some fibers still intact. Normal pressure by the weight of the structure that is attached to the ligament or capsule is being exerted. This pressure on the intact fibers can cause pain.

Procedures:

1. How to use a goniometer

A. Identify stationary arm

B. Identify movement arm

C. How to read the angle

2. Identify Passive vs. Active ROM in supine position

A. Active ROM – Client (patient) moves selected joint(s) through full ROM without any assistance. This will be the true ROM and flexibility of the tissue.

B. Passive ROM – CES moves selected joint(s) through full ROM with no assistance from client (patient)

1. Adequately stabilize clients' (patients') joints as you move them; some may have no active movement.

2. Stop when you feel the muscle restriction or the client/patient complains of pain; whichever comes first. If they are in pain, stop where they can no longer deal with the pain and then let up slightly - only 2-3 degrees and take measurement.

3. Shoulder motion

A. Flexion – ROM norms: 150°- 180°

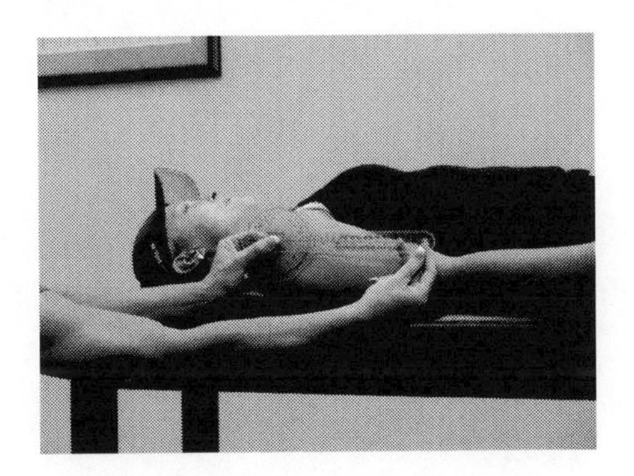

Figure 1A

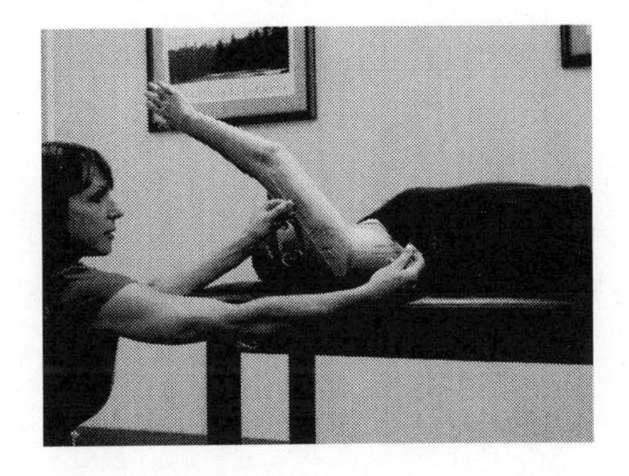

Figure 1B

1. Stationary arm – in line with body

Axis – at shoulder joint

Movement arm – in line with elbow/shaft of humerus

2. Have client in the supine position, with knees bent to flatten the lumbar spine. Arm should be extended by the sides of the body **(Figure 1A)**. This is the starting position. The arm is then raised forward and up overhead **(Figure 1B).** Palm will face in toward the body with thumb pointing upward and will end with palm still facing in, but thumb pointing down or backward. Make sure to stabilize the scapula to prevent posterior tilting, upward rotation, and elevation of the scapula. The end of ROM occurs when resistance to further motion is felt and any attempts to overcome the resistance results in posterior tilting and/or elevation of the scapula and extension of the spine.

B. Abduction - ROM norms: 150°- 180°

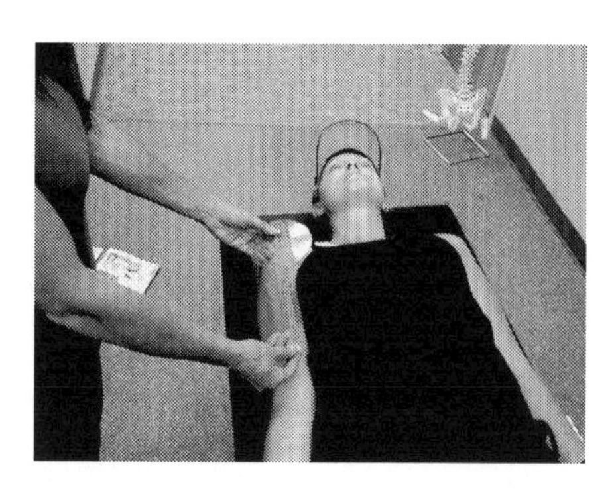

Figure 2A

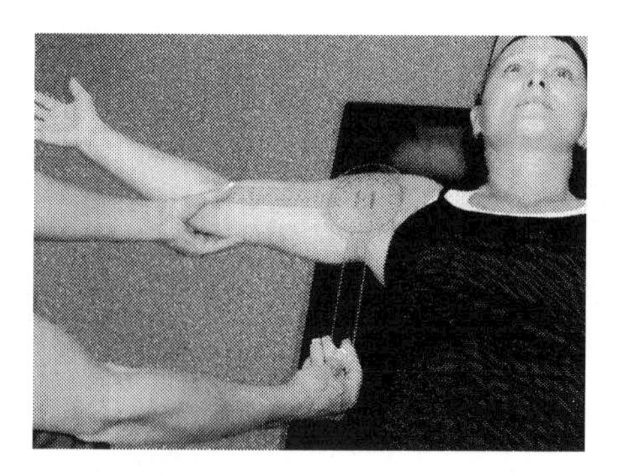

Figure 2B

1. Stationary arm – in line with body

Axis – at armpit

Movement arm – in line with elbow/shaft of humerus

2. Have client in the supine position, with knees bent to flatten the lumbar spine. Arm should be extended at the side of the body with palm facing upward **(Figure 2A).** The arm is moved laterally away from the clients trunk, moving toward the head **(Figure 2B).** The palm stays facing upward throughout the motion. Make sure to stabilize the lateral border of the scapula to prevent posterior tilting, upward rotation, and elevation of the scapula. The end of ROM occurs when resistance to further motion is felt and any attempts to overcome the resistance results in upward rotation or elevation of the scapula and/or lateral flexion of the spine.

C. Extension - ROM norms: 40°- 60°

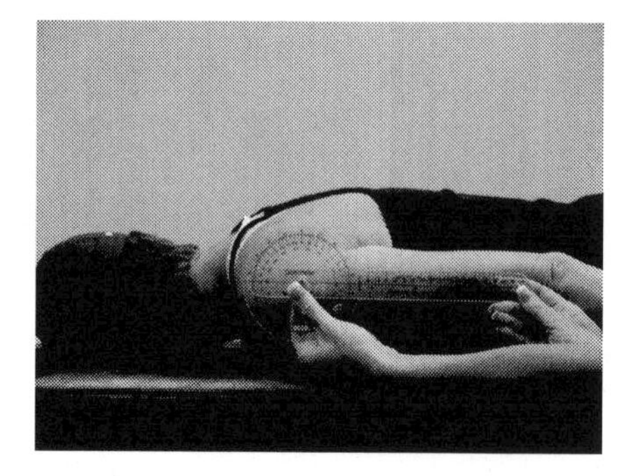

Figure 3A

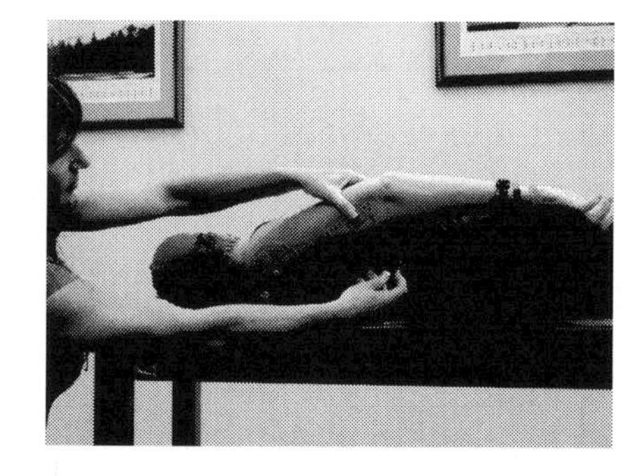

Figure 3B

1. Stationary arm – in line with body

Axis – at shoulder joint

Movement arm – in line with elbow/shaft of humerus

2. Have client lie in prone position, with face turned away from the shoulder being measured. Do not use a pillow under the head. Position the elbow in slight flexion so that it does not limit the movement **(Figure 3A).** Palm will start facing in toward the body with thumb pointing downward. Have client lift the humerus off of the table **(Figure 3B).** Make sure to stabilize the scapula at the inferior angle to prevent elevation. The end of ROM occurs when resistance to further motion is felt and attempts to overcome the resistance cause anterior tilting or elevation of the scapula and/or forward flexion or rotation of the spine.

D. External rotation - ROM norms: 70°- 90°

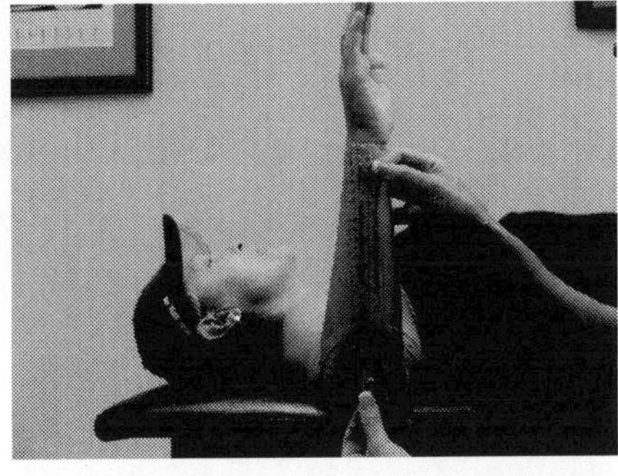

Figure 4A

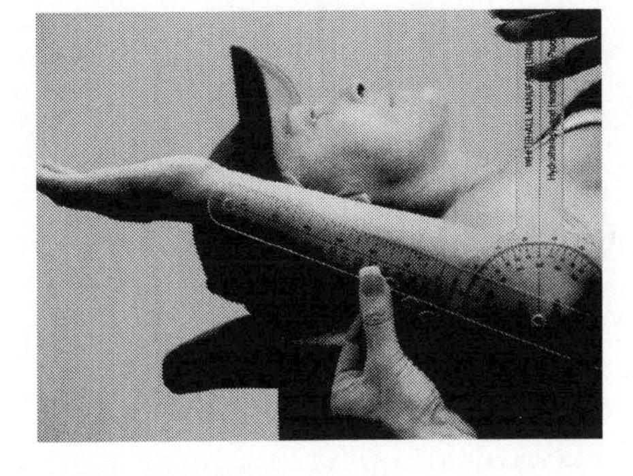

Figure 4B

1. Stationary arm – perpendicular to table or floor

Axis – elbow joint

Movement arm – along shaft of ulna (forearm)

2. Have client lie on their back with knees bent and feet flat. Have them bend their elbow to 90 degrees (at the shoulder joint). Their forearm should be perpendicular to the table or bench, with their palm facing away from them (towards feet) and their fingers pointing straight up to the ceiling **(Figure 4A).** The humerus should be supported by the table (use your knee if you are using an exercise bench) and the elbow will be unsupported. Have them rotate their forearm backward toward floor **(Figure 4B).** At the beginning of the motion, stabilize the elbow joint in order to maintain 90°. The end of the ROM occurs when resistance to further motion is felt and attempts to overcome the resistance cause a posterior tilting or retraction of the scapula and/or extension of the spine.

E. Internal rotation - ROM norms: 40°- 60°

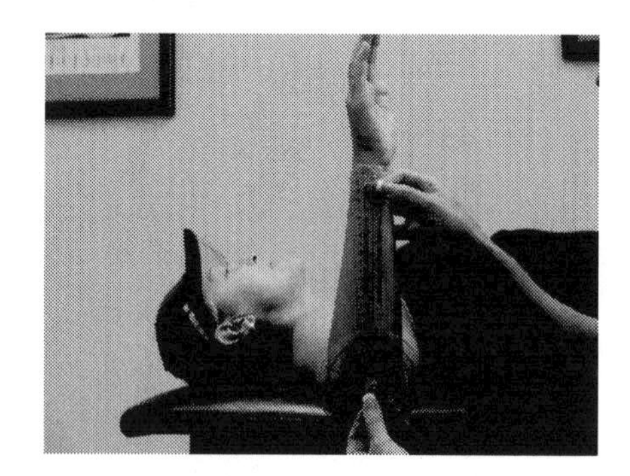

Figure 5A

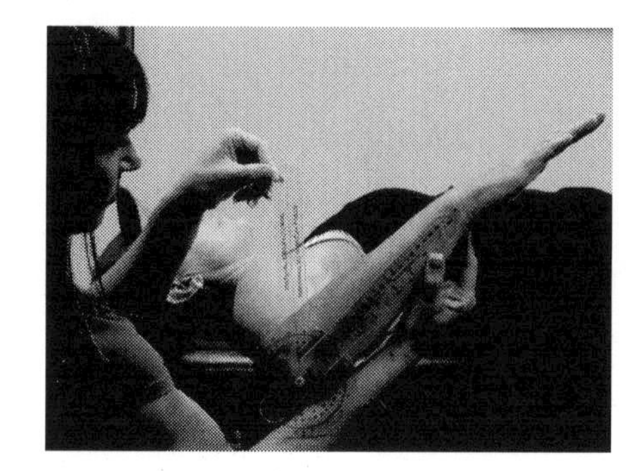

Figure 5B

1. Stationary arm – perpendicular to table or floor

   Axis – elbow joint

   Movement arm – along shaft of ulna (forearm)

2. Have client lie on their back with their knees bent and feet flat. Have them bend their elbow to 90 degrees (at shoulder joint).Their forearm should be perpendicular to the table or bench, with their palm facing away from them (towards feet) and their fingers pointing straight up to the ceiling **(Figure 5A).** The humerus should be supported by the table (use your knee if you are using an exercise bench) and the elbow will be unsupported. Have client rotate their arm forward **(Figure 5B).**Toward the end of the motion, use your hand to stabilize the clavicle and acromium processes to prevent anterior tilting and protraction of the shoulder. The end of the ROM occurs when resistance to further motion is felt and attempts to overcome the resistance cause an anterior tilting or protraction of the scapula and/or flexion of the spine.

F. Hip flexion - ROM norms: **100°-120°**

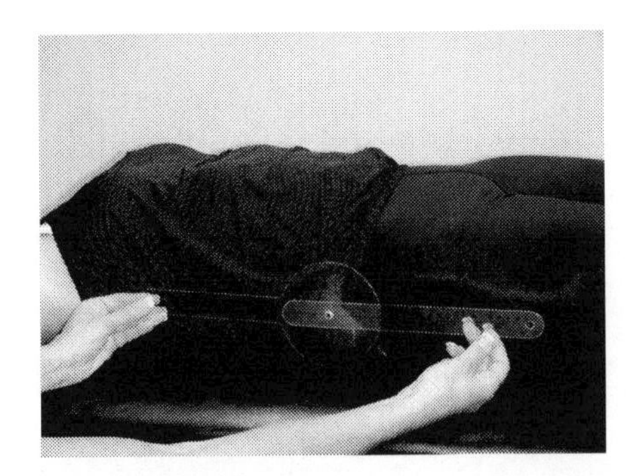

Figure 6A

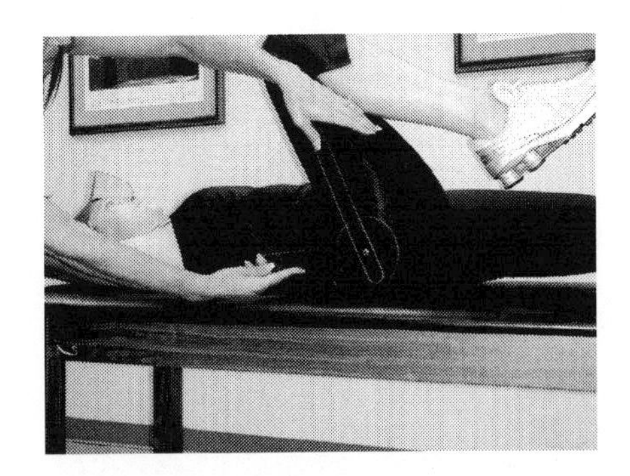

Figure 6B

1. Stationary arm – inline with the midline of the body

Axis – lateral aspect of the hip joint

Movement arm – inline with the lateral midline of the femur

2. Have the client lie on their back with both legs fully extended **(Figure 6A).** Flex the hip by lifting the thigh off of the table **(Figure 6B).** Simultaneously flex the knee. Gently move leg toward the chest. Keep the lower back flat on the table in neutral position and stabilize the pelvis with one hand to prevent posterior tilting. The end of the ROM occurs when resistance to further motion is felt and attempts to overcome the resistance cause posterior tilting of the pelvis.

F. Hip extension - ROM norms: **20°- 30°**

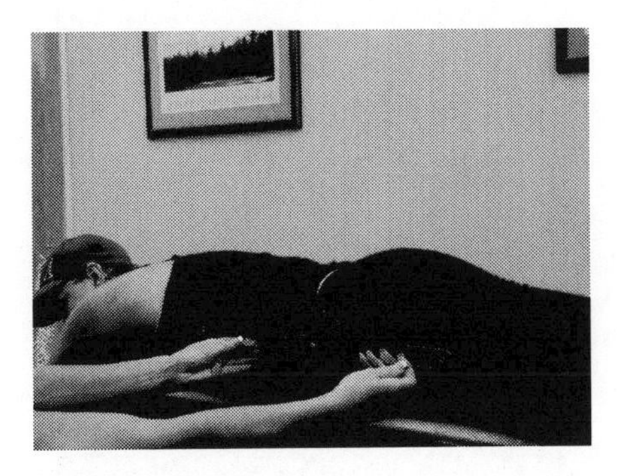

Figure 7A

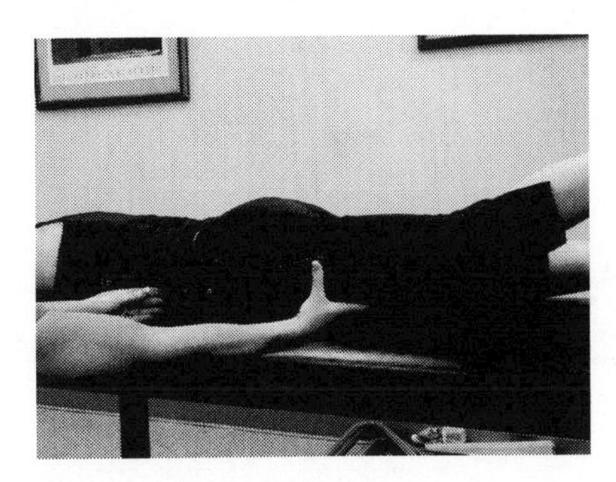

Figure 7B

1. Stationary arm – inline with the midline of the body

Axis – lateral aspect of the hip joint

Movement arm – inline with the lateral midline of the femur

2. Have the client lie in prone position with both legs fully extended and head turned to the side **(Figure 7A).** You may place a pillow under the abdomen, but do not use one under the head. Extend the hip by lifting the thigh off of the table **(Figure 7B).** Make sure that the knee is extended (not locked) throughout the entire motion. Keep the opposing leg flat on the table and hold the pelvis with one hand to prevent anterior tilting. The end of the ROM occurs when resistance to further motion is felt and attempts to overcome the resistance cause anterior tilting of the pelvis and/or extension of the spine.

F. Hip abduction - ROM norms: **40°- 42°**

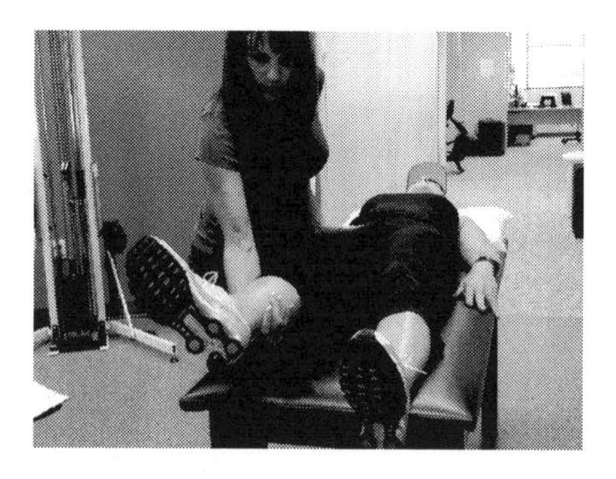

Figure 8A

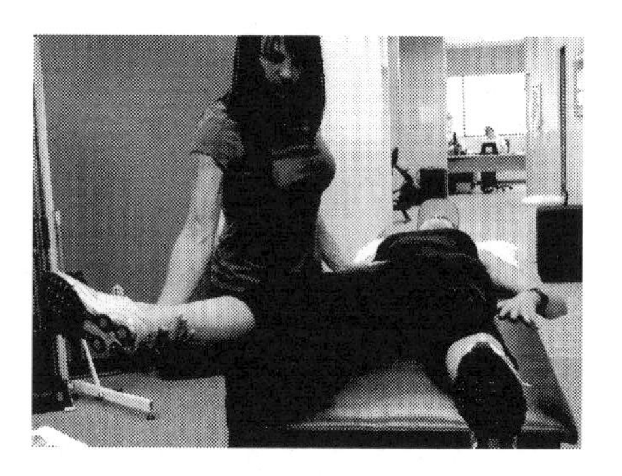

Figure 8B

1. Stationary arm – inline with and imaginary horizontal line extending across the front of the body at the level of both anterior superior iliac spines.

Axis – over the anterior superior iliac spine (ASIS)

Movement arm – inline with the anterior midline of the femur

2. Have the client lie in supine position with both legs fully extended **(Figure 8A).** Move the leg being tested laterally while keeping the opposing leg stationary **(Figure 8B).** Do not allow any rotation of the hip. Keep your hand on the pelvis at the ASIS to prevent lateral tilting, rotation, and/or lateral trunk flexion. The end of the ROM occurs when resistance to further motion is felt

and attempts to overcome the resistance cause lateral tilting, rotation, and/or flexion of the pelvis.

G. Hip adduction - ROM norms: **20°- 22°**

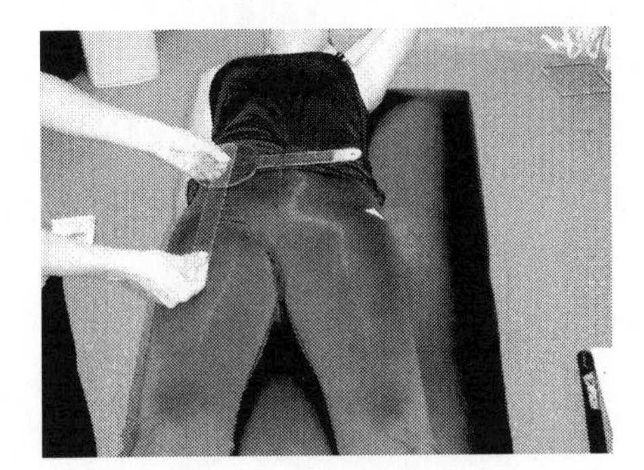

Figure 9A

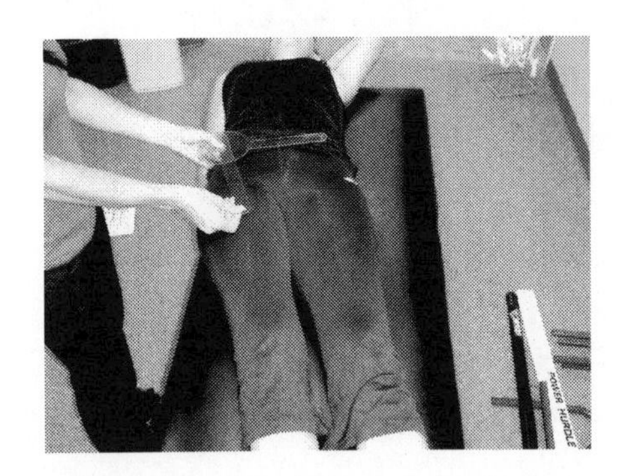

Figure 9B

1. Stationary arm – inline with and imaginary horizontal line extending across the front of the body at the level of both anterior superior iliac spines.

Axis – over the anterior superior iliac spine (ASIS)

Movement arm – inline with the anterior midline of the femur

2. Have the client lie in supine position with both legs fully extended. Abduct the opposing leg to allow for adequate movement of the leg being tested **(Figure 9A).** By placing one hand on the knee, slide the leg being tested medially toward the opposing leg. Keep one hand on the pelvis at the ASIS to keep the hip in neutral **(Figure 9B).** The end of the ROM occurs when resistance to further motion is felt and attempts to overcome the resistance cause lateral tilting, rotation, and/or flexion of the pelvis.

F. Knee flexion - ROM norms: **142°- 150°**

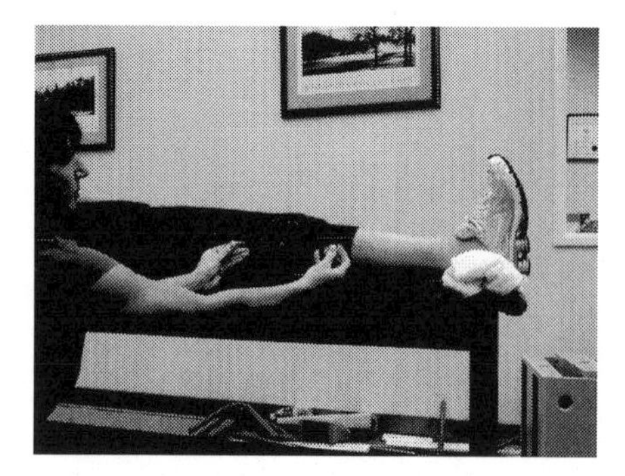

Figure 10A

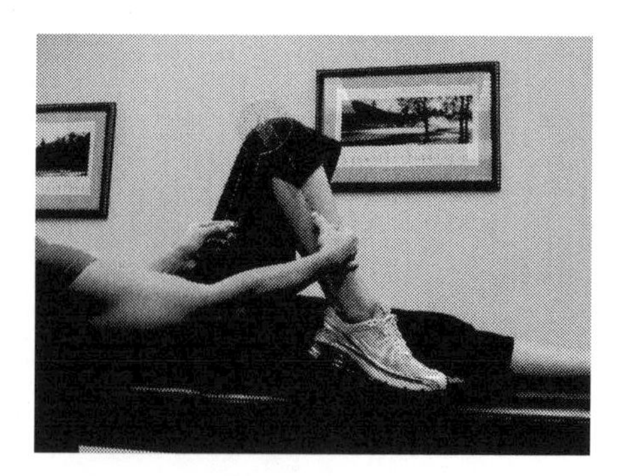

Figure 10B

1. Stationary arm – inline with the lateral midline of the femur

Axis – over the lateral epicondyle of the femur

Movement arm – inline with the lateral midline of the fibula

1. Have the client lie in supine position with both legs fully extended. Use a rolled up towel under the achilles to allow the leg to extend completely **(Figure 10A).** Do not allow any rotation, abduction or adduction of the hip. Keep your hand on your clients ankle or shin while placing their thigh at approximately 90 degrees of hip flexion. Move the knee into flexion **(Figure 10B).** The end of the ROM occurs when resistance to further motion is felt and attempts to overcome the resistance cause additional hip flexion.

F. Knee extension - ROM norms: **Return to 0 starting position of the end of knee flexion ROM**

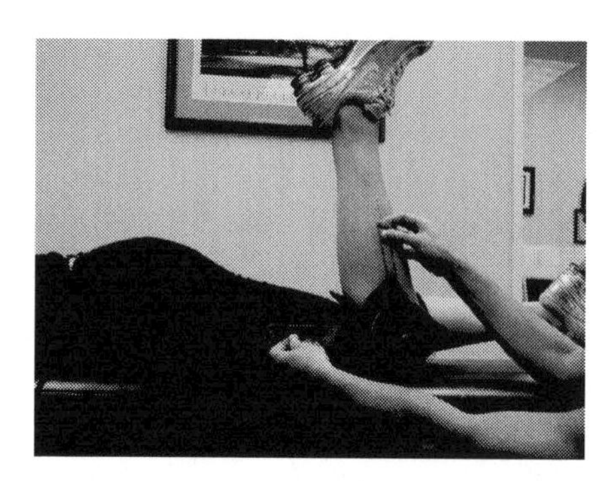

Figure 11A

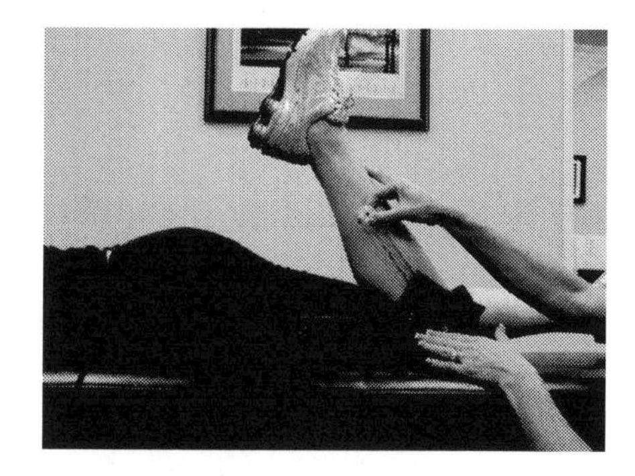

Figure 11B

1. Stationary arm – inline with the lateral midline of the femur

Axis – over the lateral epicondyle of the femur

Movement arm – inline with the lateral midline of the fibula

2. Place the client prone, with both feet off of the end of the table. Extend the knees and make sure that the hips are in neutral. Flex the knee **(Figure 11A).** Gently continue to move the knee into flexion until you feel tension from the anterior thigh and further knee flexion causes the hip to flex **(Figure 11B).**

Taking Girth Measurements

Circumference Assessment Protocol

Purpose: *to measure the circumference of the affected area during initial assessment to determine baseline measurements in order to monitor any changes in size that could indicate the onset of lymphedema*

Equipment: *tension regulated tape measure*

Procedures:

Deciding what measurements to take:

If your client has undergone a lymph node dissection or radiation to a particular area, it will be necessary to take measurements of all possible sites for lymphedema. It is important to understand the pattern of lymphatic flow (see page 243). Gravity will naturally pull the lymphatic fluid downward with the exception of the neck. If someone has lymph nodes removed or irradiated in their neck, you will want to take measurements of the neck and/or look for pitting in the jaw line and temple area. If someone has lymph nodes removed or irradiated in the chest, you will need to take measurements in the chest and waist…and so on. If, however, lymph nodes are removed or irradiated in the abdominal/pelvic area, you must take measurements at the waist and hips as well as leg measurements on both legs. Lymphedema could strike either leg. If there is significant swelling following a workout, not only do they need to see their doctor to see if they have lymphedema, it may also suggest that the intensity and duration of exercise was to intense. When your client resumes exercising, they should reduce both the intensity and duration to control swelling (this applies to clients who already have lymphedema as well). Make sure they return to you with a medical clearance form stating that it is okay for them to resume exercise with or without limitations.

A. Arm circumference

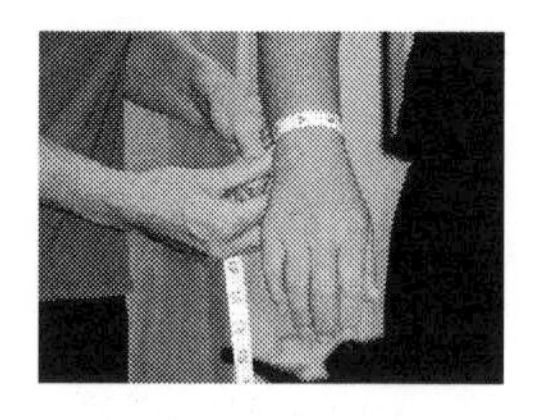

Figure 1A

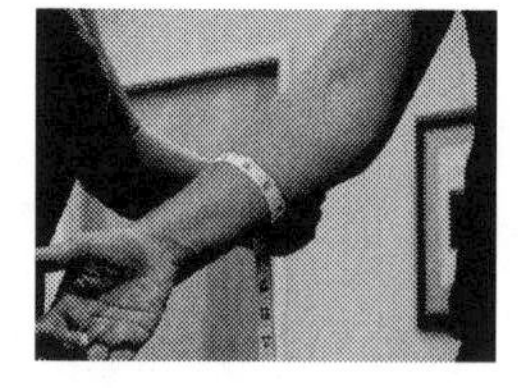

Figure 1B

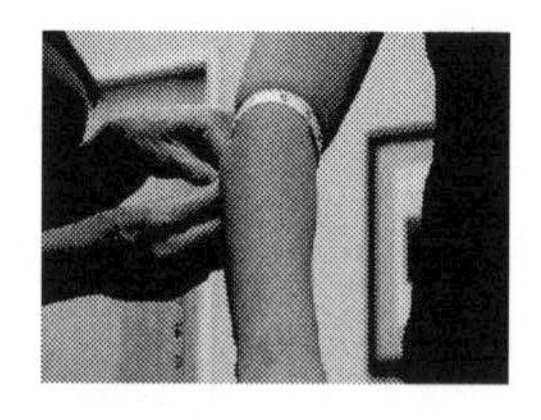

Figure 1C

Figure 1D

1. The arm circumference test is done with the participant standing erect, naturally, with the arms hanging downward but slightly away from the trunk.

2. With a tension regulated tape measure, the tester will take the following measurements:

a. Around the distal epiphysis of the ulna and radius **(Figure 1A)**

b. ½ way between the elbow crease and the epiphysis of the ulna and radius **(Figure 1B)**

c. Around the medial and lateral epicondyles of the elbow **(Figure 1C)**

d. ½ way between the elbow epicondyles and the head of the humerus **(Figure 1D)**

3. Don't pull the tape tightly; just circle the arm with it.

4. If client/patient notices fullness, redness, warmth, or pins and needles sensation in the affected area, they should perform a "pitting" test (press thumb into affected area and see if it holds the indentation). Normal tissue should blanche and refill immediately. If there is pitting, they should take measurements as well. If there is more than ½ of an inch difference from their baseline measurements, along with visible pitting, it is considered swollen and they should stop exercising and consult a physician.

B. Leg circumference

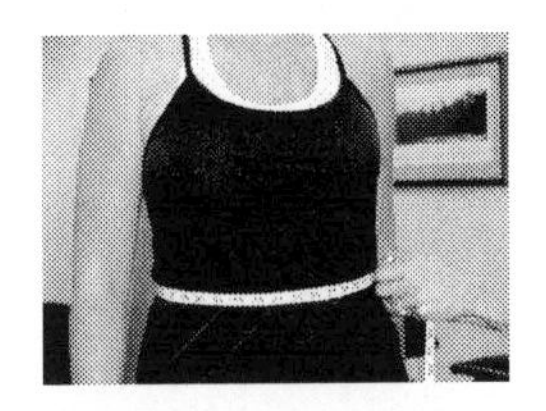

Figure 2A

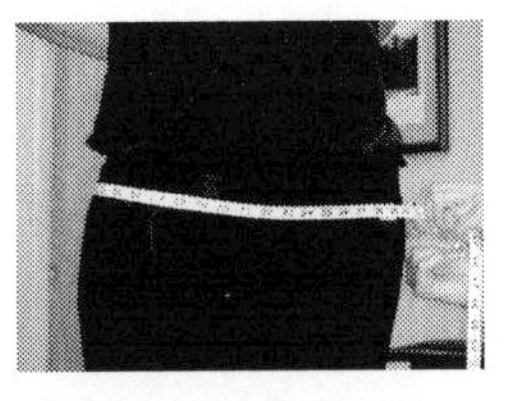

Figure 2B

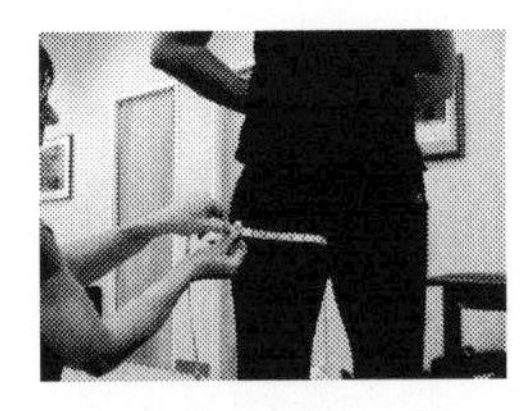

Figure 2C

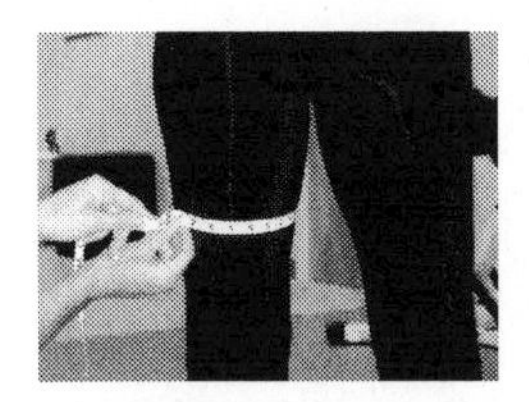

Figure 2D

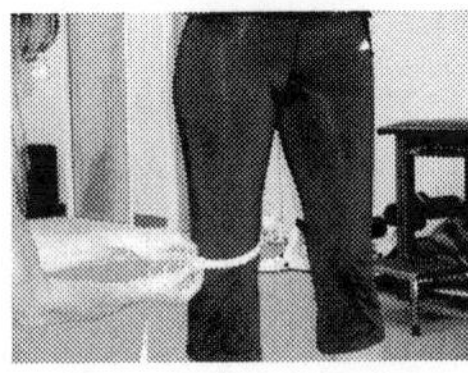

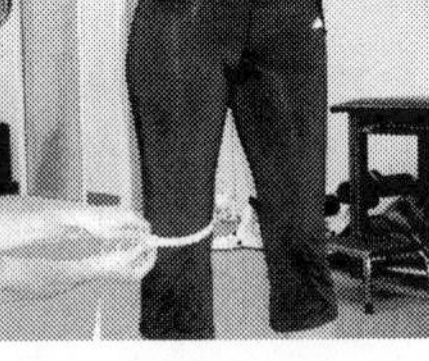

Figure 2E

Figure 2F

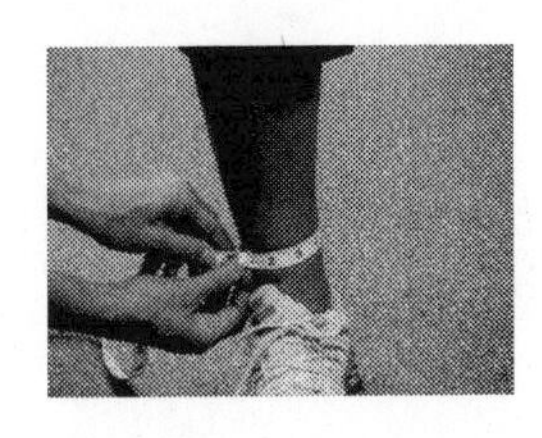

Figure 2G

1. The leg circumference test is done with the participant standing erect, naturally, with their legs slightly separated for ease of measurement.

2. With a tension regulated tape measure, the tester will take the following measurement:

a. At the waist **(Figure 2A)**

b. Across the abdomen at the belly button **(Figure 2B)**

c. At the level of the inguinal fold (bikini line, drop the tape measure horizontally) **(Figure 2C)**

d. ½ way between the inguinal fold and the knee **(Figure 2D)**

e. Around the knee **(Figure 2E)**

f. ½ way between the knee and the ankle **(Figure 2F)**

g. Around the ankle **(Figure 2G)**

3. Don't pull the tape tightly; just circle the leg with it.

4. If client/patient notices fullness, redness, warmth, or pins and needles sensation in the affected area, they should perform a "pitting" test (press thumb into affected area and see if it holds the indentation). Normal tissue should blanche and refill immediately. If there is pitting, they should take measurements as well. If there is more than ¾ of an inch difference from their baseline measurements, along with visible pitting, it is considered swollen and they should stop exercising and consult a physician.

C. Neck circumference

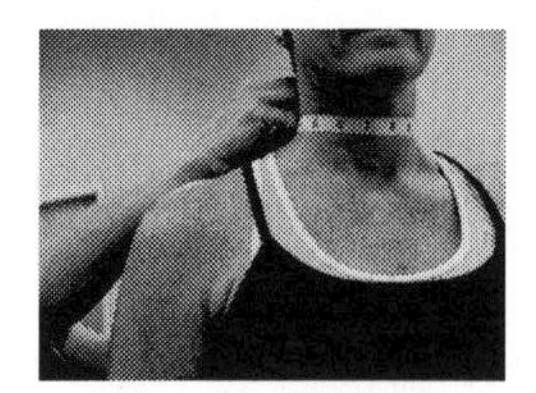

Figure 3A

1. The neck circumference test is done with the participant standing erect, naturally, with their arms by their sides.

2. With a tension regulated tape measure, the tester will take the following measurement:

a. Around the neck, just above the clavicle **(Figure 3A)**

b. Look for pitting at the jaw line and temple area

3. Don't pull the tape tightly; just circle the neck/chest with it.

4. If client/patient notices fullness, redness, warmth, or pins and needles sensation in the affected area, they should perform a "pitting" test (press thumb into affected area and see if it holds the indentation). Normal tissue should blanche and refill immediately. If there is pitting, they should take measurements as well. If there is more than ½ of an inch difference from their baseline measurements, along with visible pitting, it is considered swollen and they should stop exercising and consult a physician.

C. Abdominal/pelvic circumference

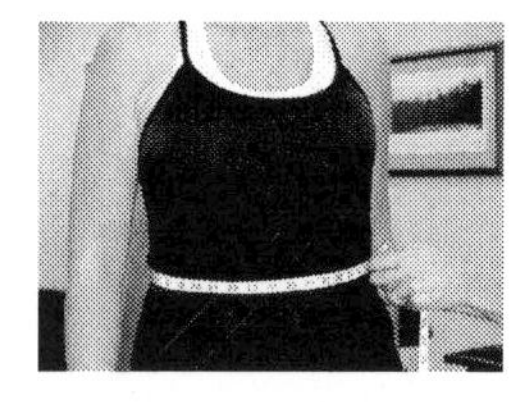

Figure 4A

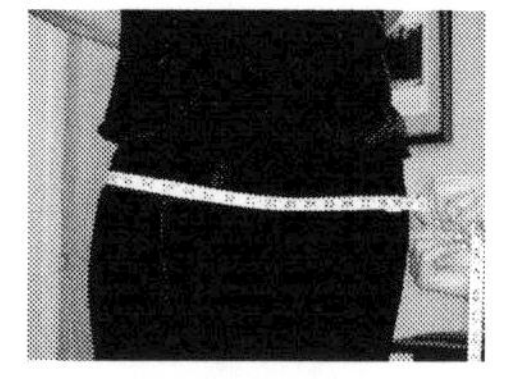

Figure 4B

1. The abdominal and pelvic circumference test is done with the participant standing erect, naturally, with their legs slightly separated for ease of movement.

2. With a tension regulated tape measure, the tester will take the following measurement:

a. Around the waist across belly button **(Figure 4A)**

b. Around the widest aspect of the hips **(Figure 4B)**

c. Take both leg measurements **(see leg circumference above – Figures 2A-2G)**

3. Don't pull the tape tightly; just circle waist and hips with it.

4. If client/patient notices fullness, redness, warmth, or pins and needles sensation in the affected area, they should perform a "pitting" test (press thumb into affected area and see if it holds the indentation). Normal tissue should blanche and refill immediately. If there is pitting, they should take measurements as well. If there is more than ½ - ¾ of an inch difference from their baseline measurements, along with visible pitting, it is considered swollen and they should stop exercising and consult a physician

Manual Stretching Techniques

A. Passive stretching - should only be utilized by the experienced fitness/health professional that is confident in their abilities to perform such stretches safely and effectively. The fitness/health professional must be careful not to overstretch their client; which could potentially lead to injuries. The client remains completely passive while the fitness professional executes the stretch. Use extreme caution when stretching the head and neck area and avoid pressing downward on the cervical spine. Stretches should be held for a minimum of fifteen seconds to get the full benefit of the stretch.

B. PNF stretching (proprioceptive neuromuscular facilitation) - should only be utilized by the experienced fitness/health professional that is confident in their abilities to perform such stretches safely and effectively. The fitness/health professional must be careful not to overstretch their client; which could potentially lead to injuries. The fitness/health professional assumes the stretch position with client/patient. The client/patient then applies force (contraction) against the fitness/health professional. This exertion is maintained for three seconds and then the client/patient is instructed to relax. As the client/patient relaxes, the fitness/health professional moves into a deeper stretch. This process is repeated three times. Make sure that you are reminding the client/patient to breath during exertion.

C. Active Isolated stretching - the client stretches themselves as far as they can - comfortably. The fitness/health professional then "assists" them to stretch a slight degree further. There should be absolutely no pain involved; only a "mild discomfort." The stretch is then held for at least fifteen seconds.

All of the stretches listed below can be done as Passive stretching, PNF stretching, or Active isolated stretching. Stretching not only relieves muscle tension and stretches through scar tissue and adhesions, it can also open up the lymphatic pathways and encourage the flow of lymph. The upper body stretches are extremely beneficial in the recovery from mastectomy and breast reconstruction. As a rule of thumb in your exercise programming, remember to stretch the areas that appear tight and to strengthen the areas that appear weak in order to maintain proper muscle balance.

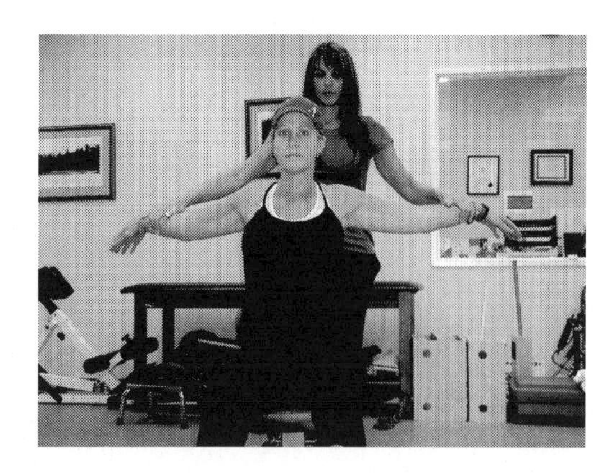

Chest Stretch One:

Support client's back by gently placing your knee against their back. Making sure that your client is erect, gently pull their wrists toward you (arm should be slightly lower than Shoulder height).

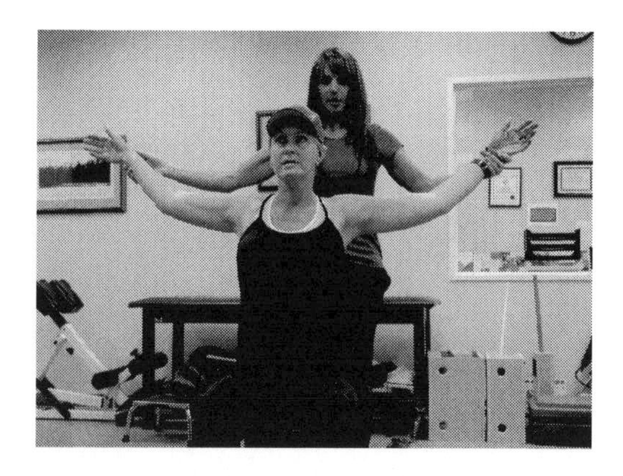

Chest Stretch Two:

Support client's back by gently placing your knee against their back. Making sure that your client is erect, gently pull their wrists toward you, up in a "V" position. This may cause some discomfort for some people. If so, skip this stretch.

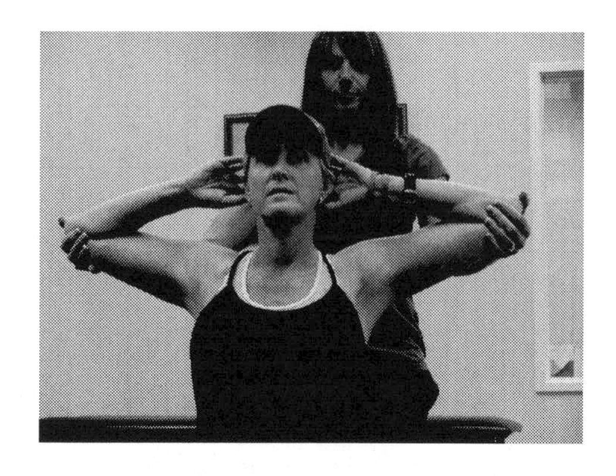

Chest Stretch Three:

Support clients back by gently placing your knee against their back. Have your client bend their elbows and place their palms against their head. Making sure that your client is erect, gently pull their elbows toward you. Do not allow their back to arch.

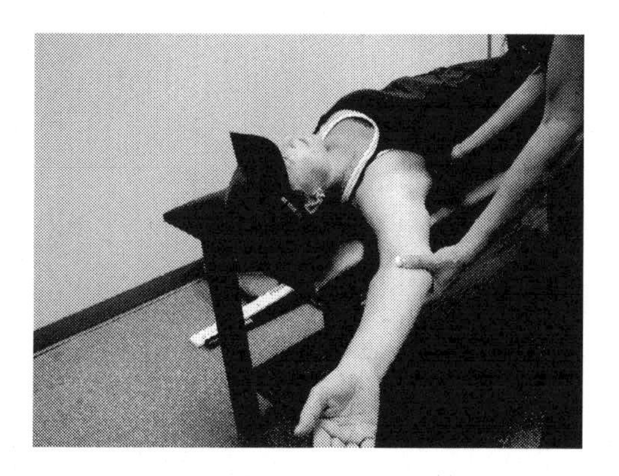

Lat Stretch:

Have your client lie on the floor with their knees bent and spine in neutral. gently "cup" their LAT and press it down toward floor – immobilizing the LAT. Have client abduct their arm (with palm up to ceiling) as you continue to apply pressure to the LAT. You can then assist them in abduction.

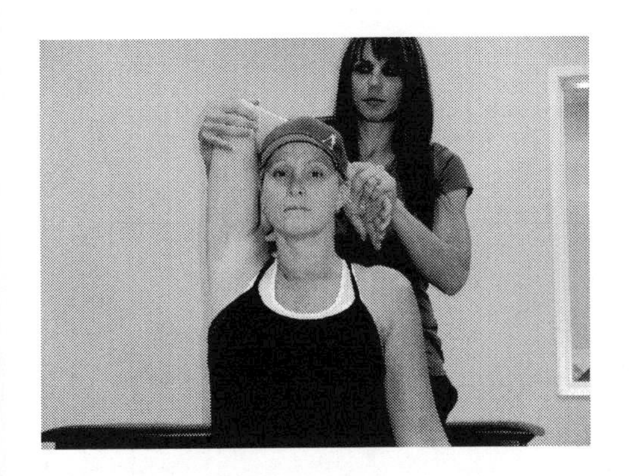

Tricep Stretch :

Support client's back by gently placing your knee against their back
Have client bend their right arm behind head with their palm facing their back. Use your right hand to gently pull their right arm toward the midline of their body while using your left hand to pull down on their right wrist. Repeat on both sides.

Shoulder Stretch:

Support client's back by gently placing your knee against their back.
Have client bend their arm and bring it across their chest with their palm facing the floor. Support their left shoulder with your left hand and gently press their right arm against their chest. Repeat on both sides. *** If this stretch causes pain in your client's shoulder, have them lower their arm below shoulder height.**

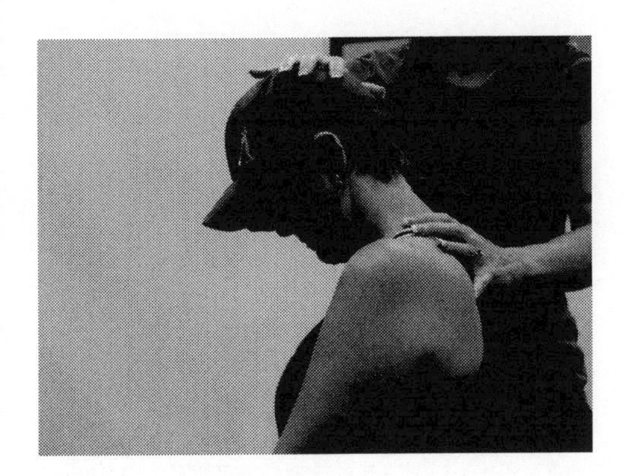

Neck Stretch One:

Place your left forearm across the top of your client's shoulders. Place the palm of your right hand on the base of your client's skull and gently press their head forward in an "arc" without pushing down-

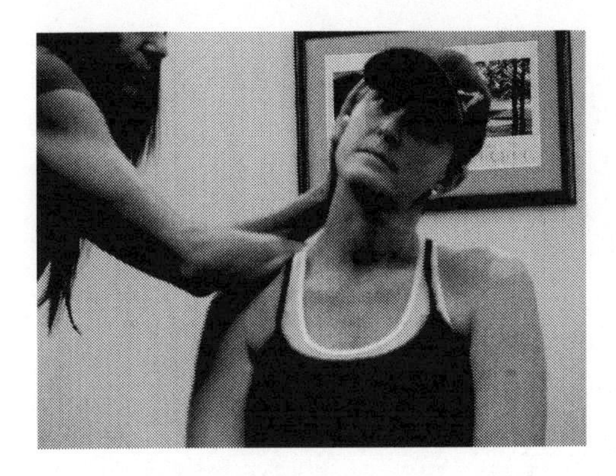

Neck Stretch Two:

Place your right forearm across the top of your client's shoulders. Place the palm of your left hand on the side of your client's head and gently press their head over toward the left shoulder while pressing downward

ward on the cervical spine. At the same time, press downward on their shoulders with your left forearm.

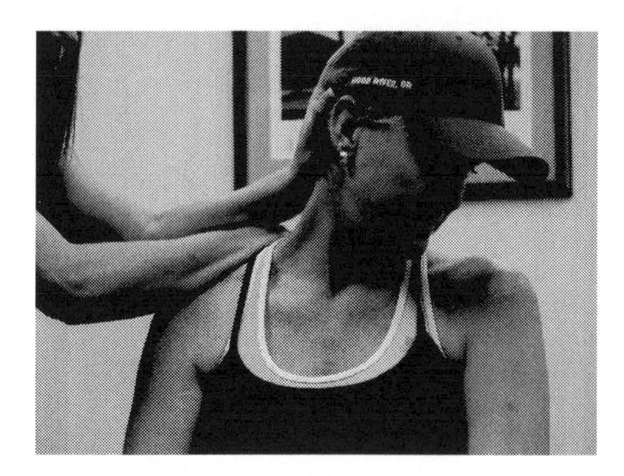

Neck Stretch Three:

Place your left hand on the base of your client's skull and gently push their head diagonally toward their left shoulder in an "arc" without pushing downward on cervical spine. At the same time, press downward on their right shoulder with your right hand. Repeat on both sides.

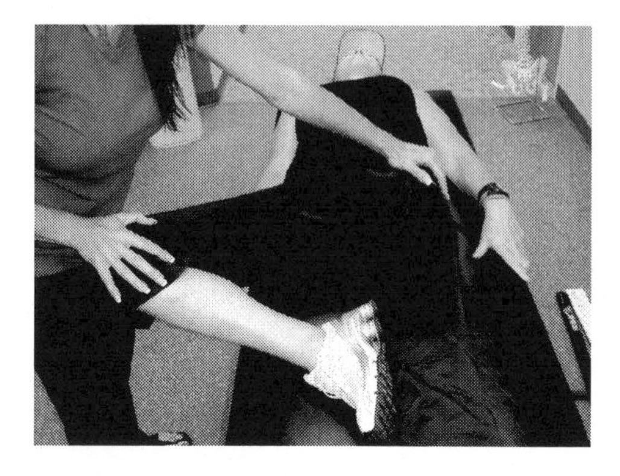

Adductor Stretch:

on the top of their shoulders with your right forearm. Repeat on both sides.

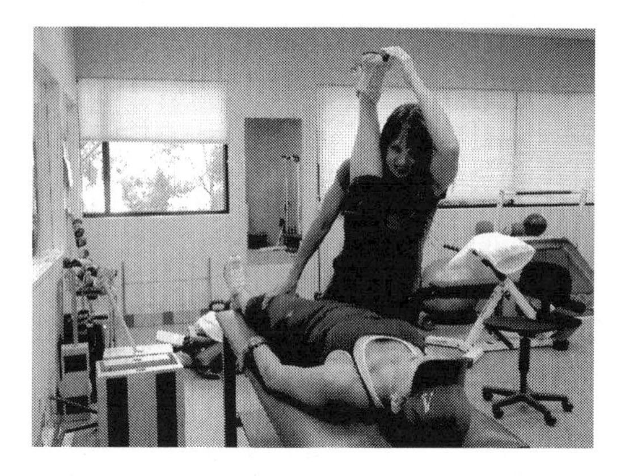

Hamstring Stretch:

Use your right leg to hold down your clients left leg (cross your shin over their ankle and use your foot to keep their foot pointing toward the ceiling). Making sure that your client does not lift their hips off of the floor, begin to straighten their right leg (keep a slight bend in their knee) by pushing on their ankle, keeping their foot flexed. Repeat on both sides.

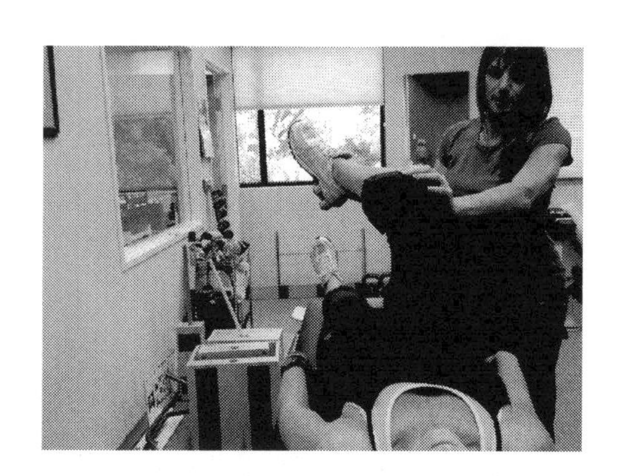

Glute Stretch:

Place your left hand on your clients' right ASIS to stabilize pelvis. Have them bend Their right leg and place the bottom of Their foot against their left thigh. Place Your right hand on the medial aspect of Their right knee and gently press downward le keeping pelvis stable with left hand. Repeat on both sides.

Place your left hand on your clients' left ASIS to stabilize pelvis. Place your right hand on their left knee (leg should be bent) and gently push knee towards opposite shoulder while keeping pelvis stable with left hand. Repeat on both sides.

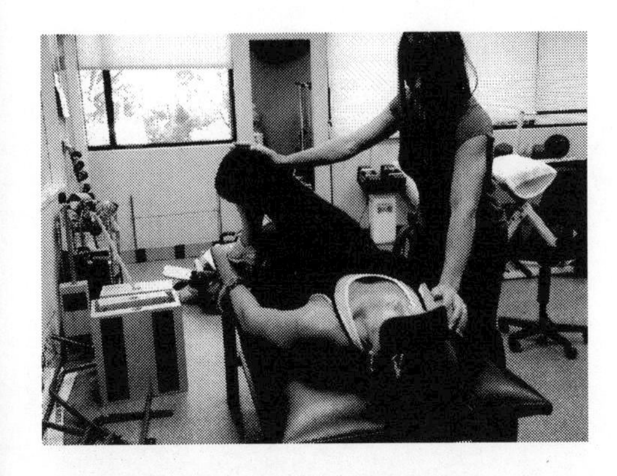

Hip and Low Back Stretch:

Use your right leg to hold down your client's left leg (cross your shin over their ankle and use your foot to keep their foot pointing toward the ceiling). Making sure that your client does not lift their hips off the floor, bend their right leg at a ninety-degree angle. Place your left hand on your client's shoulder (do not apply too much pressure on the shoulder joint). Cross their right leg over their body and place their right foot on the floor next to their left leg. Gently press their right knee toward the floor while maintaining even pressure on the shoulder. Repeat on both sides.

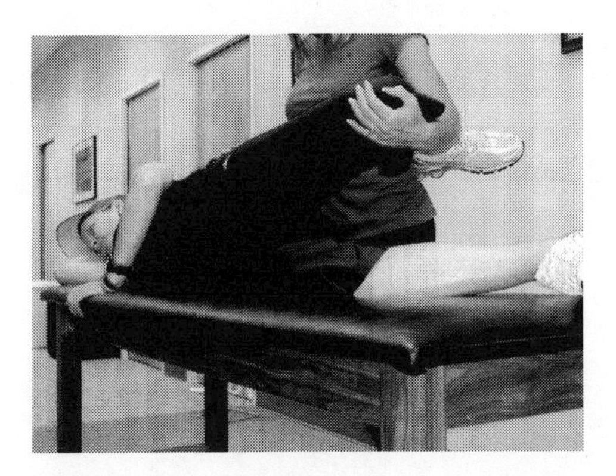

Quadricep Stretch:

Position your client on their left side. Place your left hand on their right hip and your right hand on the bottom of their right thigh (above the knee). Make sure that your client's knee is at ninety-degrees and that you don't apply any pressure to the knee joint. Have your client wrap their right foot around your left side (keeping their foot flexed. Gently pull their right thigh toward you while using your left hand to push their hip forward to keep them from rolling back toward you. Repeat on both sides.

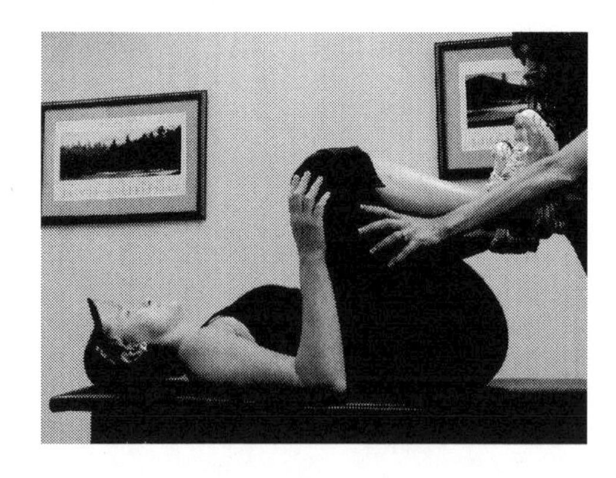

Low Back Stretch:

Have client lie on their back with their knees bent. Place your hands behind both knees and gently push their legs toward their chest while making sure that their hips don't lift off of the floor. Do not apply any pressure to the knee joints.

Exercise Recommendations

Core

Pelvic tilts on physioball:

Have client sit erect on ball (legs should be parallel to floor, or at 90 degrees. Have client engage their core and tuck their hips underneath them (bringing their ASIS and ribcage together). Pause. Then have your client tilt their hips backward (separating the ASIS and ribcage). Pause. Have them repeat 12-15x.

Hip circles on physioball:

Begin in the same position as pelvic tilts. Keeping thighs parallel and feet pointing forward, have client circle their hips clockwise 12-15x and then counterclockwise 12-15x.

Opposite arm/leg on physioball:

Have client sit erect on ball (legs should be parallel to floor). Have client engage their core and raise their right leg slightly off of the floor while reaching their left arm straight up to the ceiling. Hold for 3-5 seconds. Repeat 6x ea. Side.

Bridge set:

1) Have client lie on their back with knees bent and legs shoulder width apart. Knees should be lined up directly over ankles. With arms at their sides, have them squeeze their glutes and lift their hips off of the floor, pushing their hips toward the ceiling. Make sure that their legs do not "bow" out to the side; and that feet and thighs remain straight and parallel. Have them hold the contraction for 3-5 seconds, lower, and repeat 10-12x.

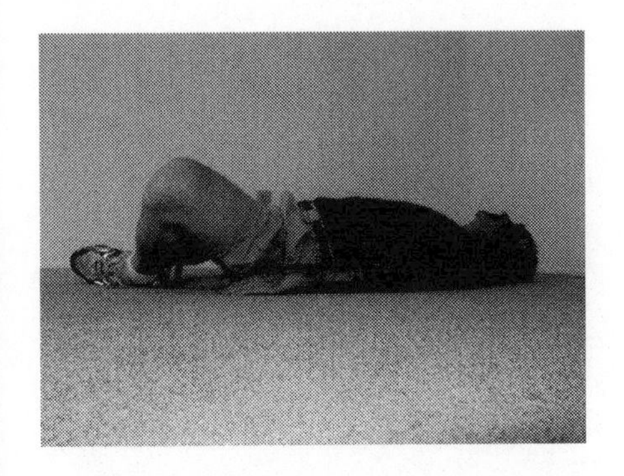

2) Have client lie on their back with knees bent out to the side and soles of feet together. With arms at their sides, have them squeeze their glutes and lift their hips off of the floor, pushing their hips toward the ceiling. Make sure that their legs remain as relaxed as possible out to the side. Have them hold the contraction for 3-5 seconds, lower, and repeat 10-12x.

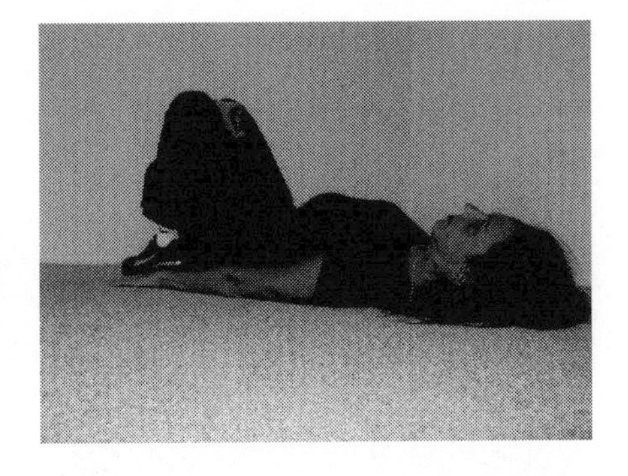

3) Have client lie on their back with knees bent and feet and legs together. Knees should be lined up directly over ankles. With arms at their sides, have them squeeze their glutes and lift their hips off of the floor, pushing their hips toward the ceiling. Have client squeeze a rolled-up towel between their legs. Have them hold the contraction for 3-5 seconds, lower, and repeat 10-12 times.

4) Have client lie on their back with knees bent and legs shoulder width apart. Knees should be lined up directly over ankles. With arms at their sides, have them lift their toes off of the floor, squeeze their glutes, and lift their hips off of the floor, pushing their hips toward the ceiling. Make sure that their legs do not "bow" out to the side; and that feet and thighs remain straight and parallel. Have them hold the contraction for 3-5 seconds, lower, and repeat 10-12 times.

Dying (dead) bug:

Have client lie on their back with their knees bent at a ninety degree angle with lower leg parallel to floor. Have them raise their arms straight up to the ceiling with the palms facing each other. Instruct them to engage their core muscles. Have them alternate raising their right arm over their head in flexion while bringing their left knee to their midline in flexion and bringing their left arm by their side while they extend their right leg. They should begin with small movements, gradually increasing their range of motion. If they experience pain in their lower back they are probably not engaging their core muscled properly; you can also have them reduce the range of motion so that they are pain-free with movement.

Marching on foam roller:

Have client lie on the foam roller with their head on one end and their low back/glutes on the other end. Have them bring their arms out to the side, palms up, slight bend in elbows, in "fly" position. Instruct them to engage their core muscles. Have them march in place on their tip toes. Make sure they are not getting excessively lordotic. If they are engaging properly, and there is no back pain, have them begin to march slowly away from their mid-section and then back again. They should repeat this until their abdominal muscles begin to get fatigued. If their arms/chest begins to get fatigued, they can cross their arms over their chest instead.

Alternating legs on foam roller:

Have client lie on the foam roller with their head on one end and their low back/glutes on the other end. Have them bring their arms out to the side, palms up, slight bend in elbows, in "fly" position. Instruct them to engage their core muscles. Have them bend both legs so that their knees are lined-up directly over their ankles. Have them raise their right leg off of the floor and parallel with the left thigh. Have them hold that position for 3-5 seconds, making sure they are not getting excessively lordotic and that there is no lower back pain. If their bent knee begins to abduct, place your hand on their inner thigh and have them gently adduct their leg against your resistance. If their leg begins to adduct, place your hand on their outer thigh and have them gently abduct their leg against your resistance. Repeat several times on each leg. If their arms/chest begins to get fatigued, they can cross their arms over their chest instead.

Balance/Core

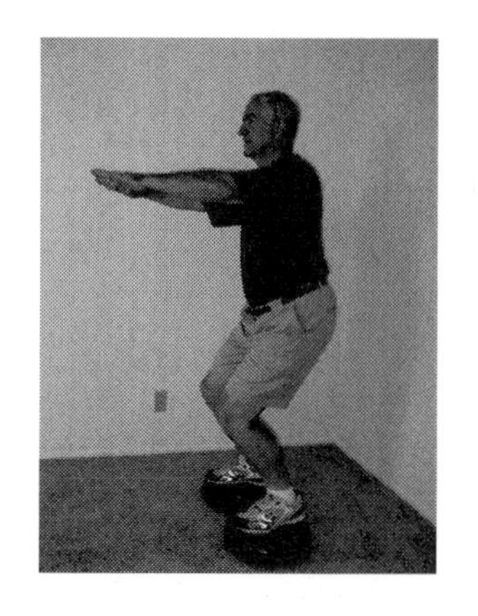

Squats on discs:

Place discs directly next to each other and have client center themselves and get "balanced." They should chose a focal point and remain focused on it the entire time. As they sit back into a squat, their arms should go out in front of them for balance. Have client squeeze their glutes as they straighten their legs (they should bring their arms to their thighs simultaneously). Repeat 12x.

Torso rotation with medicine ball on discs:

Place discs directly next to each other and have client center themselves and get "balanced." Client should engage their core and begin holding the medicine ball at shoulder height in front of their chest. They should chose a focal point and remain focused on it the entire time. Have them begin rotating their arms from side to side, keeping a slight bend in the elbows. Repeat 12x ea. direction.

Shoulder extension with medicine ball on discs:

Have client center themselves ad get "balanced." Client Should engage their core and begin holding the medicine

Ball down by their knees as they sit into a squat. As they
Come out of the squat, they should raise their arms over-
Head, keeping a slight bend in the elbows and avoiding
arching their back. Repeat 12x.

Identification and Prevention of Lymphedema

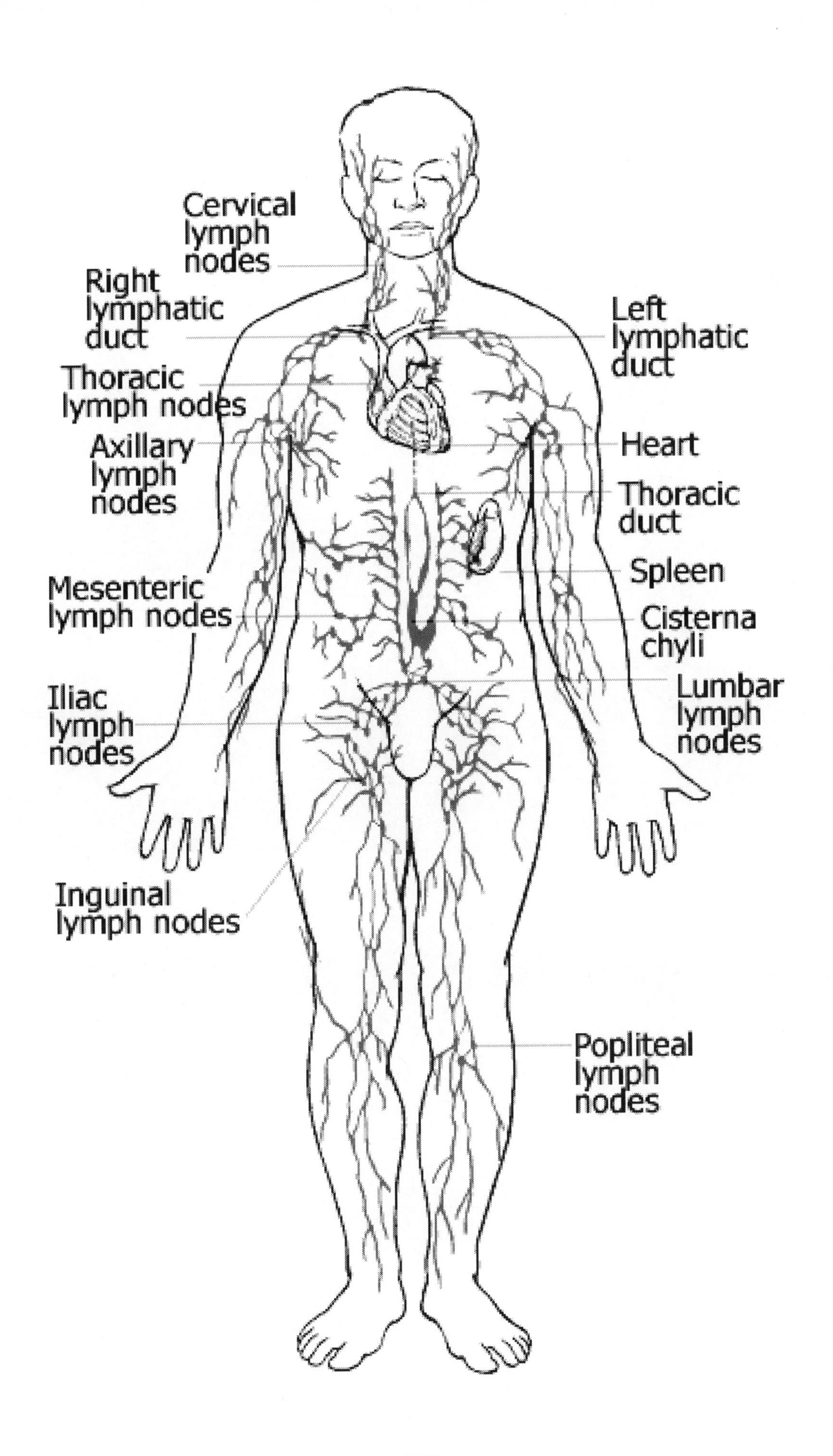
Cervical lymph nodes
Right lymphatic duct
Thoracic lymph nodes
Axillary lymph nodes
Mesenteric lymph nodes
Iliac lymph nodes
Inguinal lymph nodes
Left lymphatic duct
Heart
Thoracic duct
Spleen
Cisterna chyli
Lumbar lymph nodes
Popliteal lymph nodes

Before and after lymphedema drainage treatment::

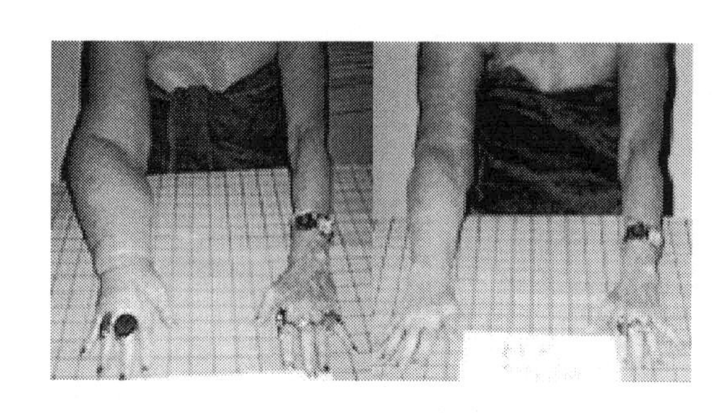

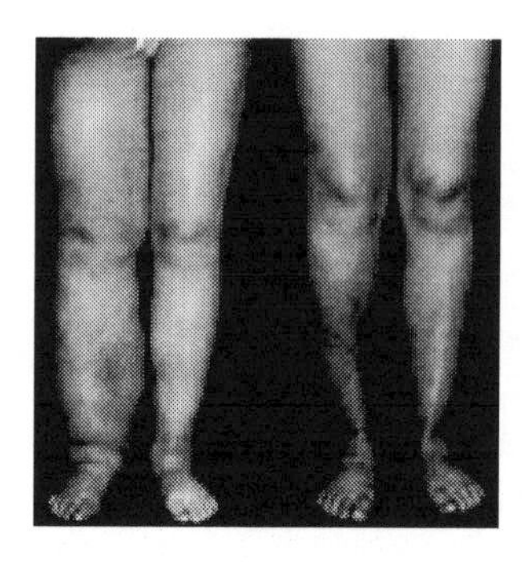

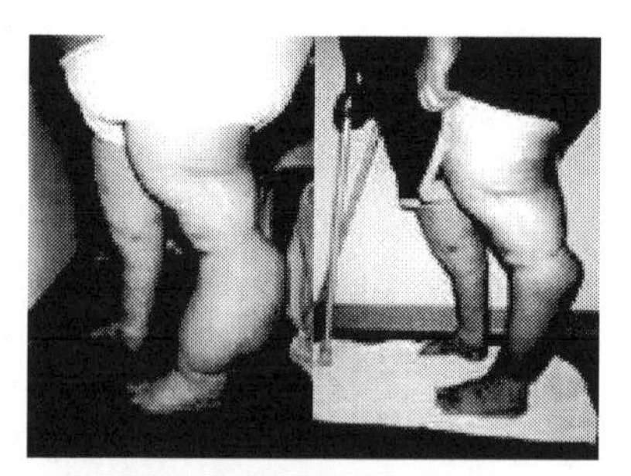

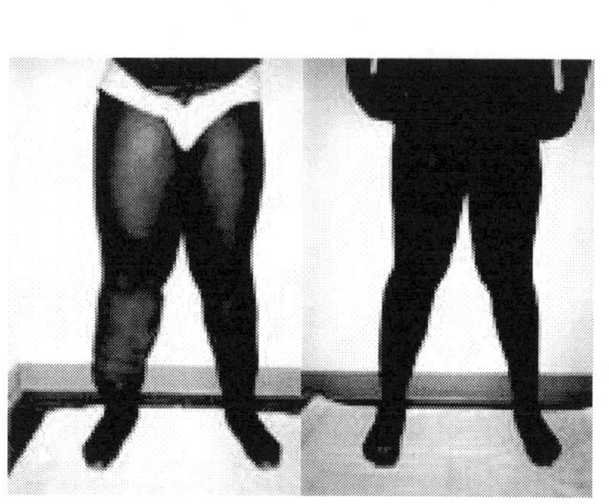

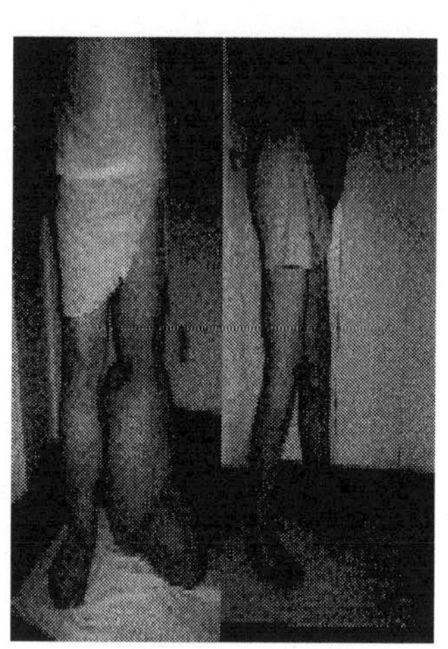

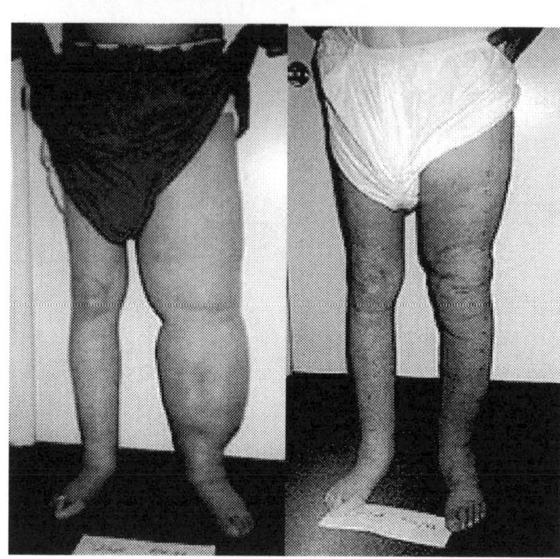

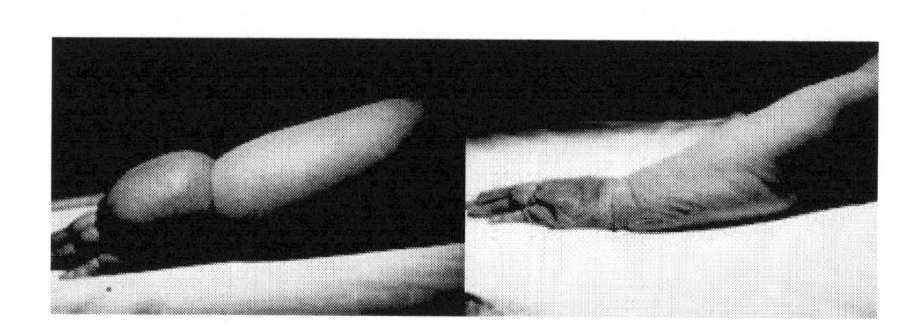

Lymphedema

The amount of lymphatic fluid that is transported through the affected areas is directly related to the amount of blood flow to those areas. Heavy lifting with the affected arm or leg, extreme climatic temperatures, extreme water temperatures when bathing, showering, or washing dishes, hot tubs, saunas, sunburn, and vigorous repetitive movements against resistance, all of which will increase blood and lymphatic flow to the affected area, should be avoided.

When a post-operative node dissection patient is fighting off an infection there will typically be an increase in lymphatic load as well as a decrease in transport capacity. Cellulitis and lymphangitis can inflame the lymphatic vessels, making them dysfunctional to transport lymphatic fluid. When patients are traveling by airplane, it is important to wear a fitted sleeve or stocking due to pressure changes which allow fluid to pool in the extremities. Due to the lack of movement during flight, the vessels which normally pump the lymph towards the regional lymph nodes are working at a very low level. Therefore, it is not only important to wear compression garments, but to move around the cabin whenever possible to prevent the pooling that can increase lymphatic load.

While each of these precautions, as well as those listed on the following pages, make perfect sense, there are several other factors that can also influence a potential lymphedema outcome. These include, but are not limited to, the number of lymph nodes that were removed from a given location, the extent of surgical disruption, the amount of lymphatic scarring from radiation, and the degree of obesity.

Upper extremity lymphedema

Lymphedema is swelling produced by an accumulation of lymph fluid in the tissue. For breast cancer patients, the swelling occurs in the arm of the affected side due to damage to the lymph vessels in the armpit area caused by the removal of the axillary lymph nodes or from radiation to that area. The job of the lymphatic vessels is to drain fluid from the tissue cells in the body, along with protein molecules, bacteria, cellular waste products, and other unusable matter. This protein-rich fluid, called lymph once it is in the lymphatic system, travels in one direction: toward the heart. It is transported through the lymphatic vessels to the lymph nodes, where it is filtered and cleansed before returning to the venous system and moving on to the heart. In the heart, the fluid is simply returned to the blood to be recirculated by the body. If the lymphatic system has been injured, as in the case of lymph node dissection or radiotherapy (for all types of cancer), the lymph can become backed up. If untreated, the backed-up fluid can provide a breeding ground for bacteria that can result in infection and can delay wound healing. A long-term accumulation of this fluid eventually results in thick and hardened tissues (fibrosis), which creates further resistance to draining the fluid from the limb. While lymphedema may not occur immediately after surgery, it can occur at any time during the rest of your life after cancer treatment. Sometimes extensive trauma can be the contributing factor while at other times it may be due to a bug bite, cat scratch, or burn. Radiotherapy also increases the chance of developing lymphedema. Radiotherapy is generally recommended to patients with a high risk of recurrence of cancer, such as those who have large, aggressive tumors. It is also recommended for those whose lymph nodes test positive for cancer cells or show an incidence of microscopic residual disease after surgery. In most cases when breast conserving surgery is performed, radiotherapy is given to the rest of the breast tissue. Even after a modified radical mastectomy, radiotherapy is recommended if a patient is at high risk. Because lymph nodes are radiosensitive, radiotherapy depletes the lymphocytes in the nodes and decreases their filtering function and immune function. Lymphedema can only affect the arm where the nodes have been irradiated, not the overall immune system. After radiotherapy, the nodes become scarred and fibrotic, increasing the potential for blockage. Upwards of five percent of breast cancer patients are afflicted with lymphedema during their first year after surgery. The lifelong rate for affliction is between 8-30%. Although this example if pertaining to breast cancer and a axillary node dissection, it is important to remember that if you have undergone a lymph node dissection or radiation for any type of cancer, you are at risk for lymphedema in that part of your body. With proper education and care, lymphedema can be avoided, or, if it develops, kept well under control. Older individuals and those with poor nutrition face an increased risk, as do individuals with infections. Removal of the nodes and damage to the area prevent the lymph fluid in the arm from draining properly, allowing it to accumulate in the tissue by restricting pathways and causing back-up. It has been well documented that the development of lymphedema after breast cancer surgery and radiotherapy is related to the extent of the lymph node dissection, the extent of the breast surgery, and whether radiotherapy is given to the axilla. When breast conservation surgery is performed without radiotherapy or an axillary node dissection, there is no incidence of lymphedema. If lymphedema goes untreated, it can result in decreased arm function, decreased range of motion in the arm and shoulder, decreased finger function and numbness in the hand, and swelling of the entire arm, hand, and fingers. In addition, this damage may result in pain and tightness in the area as the lymph vessels close up, tighten, and sometimes snap. Guarding against infection is extremely important because the affected arm will be more susceptible to infection than the uninfected arm, and infection can cause increased swelling. If you notice any signs of infection, contact your doctor immediately. These signs include swelling, fever, or skin that is red, tender, warm, persistently itchy or blotchy. There are three categories for grading lymphedema; grade one, two, and three. In grade one, when the skin is pressed the pressure will leave a pit that takes some time to fill back in. This is referred to as *pitting edema*. Sometimes the swelling can be reduced by elevating the limb for a few hours. There is little or no fibrosis at this stage, so it is usually reversible. In grade two, when the swollen area is pressed, it does not pit, and the swelling is not reduced very much by elevation. If left untreated, the tissue of the limb gradually hardens and becomes fibrotic. In grade three, the lymphedema is often referred to as *elephantiasis*. It occurs almost exclusively in the legs after progressive, long-term, and untreated lymphedema. At this stage there may be gross changes to the skin. There may even be some leakage of fluid through the tissue in the effected area, especially if there is a cut or sore. While lymphedema will respond to treatment, at this stage it is rarely reversible.

Lymphedema is a very serious condition and should not be taken lightly. It frequently results in complications, such as lymphangitis (a bacterial infection of the lymphatic system), skin changes, fibrosis, and infection. There are even a few life-threatening complications, although rare, such as the development of a rare type of cancer, lymphangiosarcoma, in the affected area. This can occur in patients with long-term, untreated, or improperly treated lymphedema. Unfortunately, this condition requires immediate amputation. Lymphedema may worsen with time if it is not attended to. It can become disabling by stiffening the joints or making the limbs heavy, and may cause significant cosmetic deformities.

Using compression bandages

Compression bandages apply external pressure to a swollen limb. When swelling has persisted in an area, the tissue loses some of it's elasticity and does not return to its' original position and shape, even when the fluid decreases. The bandages support the skin and underlying blood vessels. Bandaging usually starts with gauze tape at the fingers and then continues with a series of different sized short-stretch bandages around the hand, progressing up the arm to within a short distance of the shoulder. The number of bandages used depends on the size of the arm and how effectively the compression is achieved. Many therapists recommend wearing the bandages while sleeping, as well as anytime that you are engaging in physical activity or exercising. I addition to bandaging, most patients should be fitted with a compression sleeve that is worn during the day time. The garments are not designed to reduce swelling, however, but to maintain the size of the limb and prevent swelling from increasing. Some patients will use a sleeve when flying on an airplane in order to reduce their chances of getting lymphedema due to the changes in cabin pressure.

Exercise and lymphedema

One of the most important, and often over looked component of exercise is breathing. Not only does breathing allow precious oxygen to be circulated through the bloodstream, but it is also effective for moving fluid through a gentle pumping action of the abdominal muscles. The fluid is pumped through the central lymphatic vessel in the chest cavity, stimulating the flow of lymph. When you breathe in, using your abdominal muscles, the pressure in the chest cavity changes, because the belly breath moves your diaphragm. When you exhale, the pressure changes once again. This back-and-forth alternation in the pressure acts like a pump on the large lymphatic trunk that runs up through the chest cavity and drains into the venous system of the neck.

Here's how; sit in an upright position. Take a deep breath through your nose and exhale through your mouth, flattening your belly and squeezing out every last bit of air. Emptying the lungs completely and removing all of the stale air from the bottom of the lungs automatically stimulates a diaphragmatic breath. Breathe in through your nose and notice how your belly expands. Repeat the sequence again. Let the air out through your mouth, making sure your belly flattens. Try another one or two breaths this way. If you get light-headed, try to slow down your inhalation, and pause before breathing in again. It is not necessary to breath with a giant breath – just one that goes to the bottom of your lungs, while your chest remains still. Imagine a balloon in your stomach that inflates when you inhale and deflates when you exhale.

Everyone knows the benefits of exercise for seemingly healthy individuals, but they carry even more a punch for those suffering from lymphedema. Generally speaking, those who exercise have a lower percentage of body fat. Keeping body fat in check can actually help to prevent lymphedema. Fat can be a special problem with an impaired lymphatic system. Fat is deposited in the interstitial tissue and can make it more difficult for the fluid to pass through and into the lymph vessels. The lymphatic system is stimulated by the pumping action of the blood vessels, as well as the pumping action of muscles, so anything one does to improve their circulatory system will be helpful for the lymphatic system. A good exercise goal, for aerobic activity, is thirty minutes three or four times a week. Choose an exercise that will allow your client to mildly increase their heart rate without "over-doing" it. It is important that they wear their support garment or bandage while exercising. Bandages increase pressure against the skin during exercise. The pressure, coupled with the contraction of their muscles, encourages the lymph to move. Exercise, in and of itself, will help to pump the muscles, improve circulation, and move lymph from congested areas into an area where it can drain more efficiently. It is important to start out very slowly with few repetitions and wait until the next day to see how the affected limb has responded. You can gradually increase their repetitions, based on what the limb will tolerate. Have clients take their time and use good form and posture. Clients should not exercise under the following conditions:

- When they have a fever
- If they experience chest pain
- If they experience sudden shortness of breath or unusual fatigue
- When they have recurring leg pain or cramps
- If they experience an acute onset of nausea during exercise
- If they feel disoriented or confused
- When they have had recent bone, back, or neck pain that is not relieved with rest

- If they have an irregular heartbeat

Stages of Lymphedema

- **Stage 1** (mild) - When you awake in the morning, your arm may be a normal size. The tissue is still in a "pitting stage" (when pressed by a finger the area indents and holds the indentation). This is called "Pitting Edema."

- **Stage 2** (moderate) - The tissue does not "pit" (when pressed by a finger the tissue bounces back without leaving an indentation). The tissue has a spongy consistency.

- **Stage 3** (severe) - The tissue at this stage is hard (fibrotic) and will be unresponsive to the touch. The **swelling is irreversible** and the limb is **very large and swollen**.

- Infections are possible at any stage of lymphedema but occurrence becomes greater as stages progress. A swollen limb, left untreated, becomes hard (fibrotic) and full of lymph fluid which is high in protein and a perfect medium for bacteria and infections

Precautions to avoid upper extremity lymphedema

- Avoid insect bites, burns, skin irritants, hangnails, and torn cuticles (wearing gloves while doing housework or gardening is a great idea)
- Avoid tight fitting jewelry on the affected arm or hand
- Wear loose fitting clothing on arms, chest, and shoulders
- Don't overheat – avoid saunas, whirlpools, steam rooms, hot baths, and sun bathing
- Don't receive shots, have blood drawn, or have blood pressure taken on affected arm
- If you are overweight and have experienced swelling, losing weight can help reduce it by reducing the amount of fatty tissue which retains fluid and blocks lymphatic pathways
- Remember that tennis, racquetball, golf, and bowling are al considered risky sports
- Keep the at-risk arm(s) spotlessly clean and use lotion after bathing
- Avoid repetitive movements such as scrubbing, pushing, or pulling, with the at-risk arm
- Avoid heavy lifting with the affected or at-risk arm. Never carry heavy handbags or bags with over-the-shoulder straps on the affected arm
- When traveling by air, patients with lymphedema, or those who are at risk, must wear a well-fitted compression sleeve
- Use an electric razor when shaving neck (cervical lymph nodes), armpits (axillary node dissection), or legs (lower extremity lymph node dissection)

Lymph Drainage Exercises for Upper Extremity Lymphedema

Prevention and Management

Prior to beginning these exercises, clients should start with a five-minute aerobic warm-up to get the juices flowing. As they begin each of the following exercises make sure to take several deep abdominal breaths (as were described earlier).

1.) **Pelvic tilt** – have client lie on their back with their knees bent and feet flat on the floor. Have them tilt their hips so that they are able to press the small of their back against the floor. Have them pause for several seconds then release the contraction. Repeat five times.

2.) **Modified sit-up** – have client lie on their back with their knees bent and feet flat on the floor. Have them perform a pelvic tilt, pressing the small of their back to the floor. Have them keep their neck in neutral and their chin pointing to the ceiling. As they exhale, have them lift up their chest and shoulders, pausing when they feel their abdominal muscles tighten up. Have them slowly lower themselves back to starting position (trying not to rest between repetitions). Repeat as many times as they can comfortably.

3.) **Neck stretches** – have client begin by standing or sitting erect. Have them exhale and turn their head slowly to the right, looking over their shoulder. Have them inhale as they return to center. Have them repeat this to the left. Next, have them tilt their head to the right, allowing their chin to drop toward their shoulder. Have them maintain this position for five seconds, breathing regularly. Slowly have them bring their head back to center. Have them repeat this to the left. Finally, have them tilt their head to the right, allowing their ear to drop toward their shoulder. Have them maintain this position for five seconds, breathing regularly. Slowly have them bring their head back to center. Have them repeat this to the left.

4.) **Shoulder shrugs** – have client shrug both shoulders, lifting them towards their ears as they inhale. Have them exhale, and return to a relaxed position. Next, have them exhale and press their shoulders down as far as possible, pause. Have them inhale and return to the relaxed position.

5.) **Shoulder rolls** –have your client lift their shoulders up to the ears then rotate the shoulders back and down, making a smooth, continuous motion. Repeat several times. Repeat in the other direction.

6.) **Isometric shoulder blade squeeze** – have your client bend their elbows to a right angle, parallel to the floor. Have them exhale and pull them towards the center of your back, squeezing the shoulder blades together. Pause. Have them inhale and return to starting position. Repeat 8-12 times.

7.) **Isometric chest press** – have client place the palms of their hands together, with their elbows bent and arms parallel to the floor at shoulder level. Have them exhale and push their hands firmly together. Pause. Have them inhale and relax. Repeat 8-12 times.

8.) **Shoulder circles** – have your client hold their arms at shoulder height, parallel to the floor, with their palms facing down. Have them exhale and rotate their arms so that their palms are facing upward. Have them inhale and return palms to starting position. Repeat 8-12 times.

9.) **Wrist circles** – have client rotate their fist in small circles, isolating the movement to the wrist only. Rotate several times in one direction, then in the other.

10.) **Wrist flexion and extension –** have your client bend their wrist towards them, then away from them, isolating the movement to the wrist only. Repeat several times

11.) **Fist clench** – have your client open their hands and stretch their fingers, spreading them apart. Then have them slowly clench each hand to make a fist. Hold for five seconds, breathing regularly, then relax. Repeat 8-12 times.

Lower extremity lymphedema

At risk for lower extremity lymphedema is anyone who has had gynecological, melanoma, prostate or kidney cancer in combination with inguinal node dissection and/or radiation therapy. Just as with upper extremity lymphedema, it is caused by an accumulation of lymph fluid in the tissue, can occur at any time after surgery, and can be prevented and/or controlled with proper care. Lower extremity lymphedema may cause swelling in the toes, foot, ankle, leg, abdomen, and genitals. It is crucial to notify your doctor immediately if you notice an increase in swelling in any of these areas. Older individuals and those with poor nutrition face an increased risk, as do individuals with infections. Removal of the nodes and damage to the area prevent the lymph fluid in the from draining properly, allowing it to accumulate in the tissue by restricting pathways and causing back-up. In addition, this damage may result in pain and tightness in the area as the lymph vessels close up, tighten, and sometimes snap. Guarding against infection is extremely important because the affected area will be more susceptible to infection , and infection can cause increased swelling. If you notice any signs of infection, contact your doctor immediately. These signs include swelling, fever, or skin that is red, tender, warm, persistently itchy or blotchy.

Precautions to avoid lower extremity lymphedema

- Avoid insect bites, burns, skin irritants, and cutting cuticles(be cautious during pedicures)
- Don't overheat – avoid saunas, whirlpools, steam rooms, hot baths, and sun bathing
- If you are overweight and have experienced swelling, losing weight can help reduce it by reducing the amount of fatty tissue which retains fluid and blocks lymphatic pathways
- Use an electric razor to remove hair from the affected leg
- Never allow an injection or a blood drawing in the affected leg(s)
- Keep the at-risk leg spotlessly clean and apply lotion after bathing
- Avoid vigorous repetitive movements against resistance with the affected leg(s)
- Do not wear socks, stockings, or undergarments with tight elastic bands that can restrict blood flow
- When traveling by air, patients with lymphedema, or those who are at risk, must wear a well-fitted compression stocking
- Use an electric razor when shaving legs

Lymph Drainage Exercises for Lower Extremity Lymphedema

Prevention and Management

Prior to beginning these exercises, clients should start with a five-minute aerobic warm-up to get the juices flowing. As they begin each of the following exercises make sure to take several deep abdominal breaths (as were described earlier).

1) **Pelvic tilt** – have client lie on their back with their knees bent and feet flat on the floor. Have them tilt their hips so that they are able to press the small of their back against the floor. Have them pause for several seconds then release the contraction. Repeat five times.

2) **Modified sit-up** – have client lie on their back with their knees bent and feet flat on the floor. Have them perform a pelvic tilt, pressing the small of their back to the floor. Have them keep their neck in neutral and their chin pointing to the ceiling. As they exhale, have them lift up their chest and shoulders, pausing when they feel their abdominal muscles tighten up. Have them slowly lower themselves back to starting position (trying not to rest between repetitions). Repeat as many times as they can comfortably.

3) **Bicycles** – have client lie on their back with their knees bent at ninety degrees (shins are parallel to the floor). Have them perform a pelvic tilt, pressing the small of their back to the floor. Keeping their upper body relaxed on the floor, they should begin pumping their legs back and forth, in very small motions, like a bicycle. It is very important not to allow them to extend their legs too far out, as that will put excessive pressure on their lumbar spine.

4) **Leg circles** – have client lie on their back with left knee bent and foot on floor and right leg extended up toward the ceiling. Have client perform clockwise hip circumduction, using very small and controlled movements 5-6 times. Have them repeat going counterclockwise. Have them repeat the entire process on the other leg.

5) **Knee flexion/extension** – have client lie on their back with left knee bent and foot on floor and right leg extended up toward the ceiling. Have them bend and extend the right leg 8-12 times. Have them lower the right leg, bend it, and place the foot on the floor. Have them extend the left leg up to the ceiling and repeat the process on that side.

6) **Plantar/dorsi flexion** - have client lie on their back with left knee bent and foot on floor and right leg extended up toward the ceiling. Have client point and flex their right foot at the ankle joint 8-12 times. Have them lower the right leg, bend it, and place the foot on the floor. Have them extend the left leg up to the ceiling and repeat the process on that side.

7) **Foot circles** - have client lie on their back with left knee bent and foot on floor and right leg extended up toward the ceiling. Have client perform clockwise circumduction with the right foot 8-12 times. Have them repeat going counterclockwise. Have them lower the right leg, bend it, and place the foot on the floor. Have them extend the left leg up to the ceiling and repeat the process on that side.

Cancer Related Pain

Cancer Related Pain

Not all people with cancer experience pain and not all cancers produce pain equally. Some cancers, even when advanced, may not produce pain at all. Cancers involving bone, either directly or through the spread of the disease are usually associated with pain when advanced. Pain can have a terrible effect on one's quality of life and ability to function. It can lead to depression, irritability, withdrawal from social activity, anger, loss of sleep, loss of appetite, and an inability to cope. Pain may be acute or chronic. Acute pain is severe and lasts a relatively short time. It is usually a signal that the body is being injured in some way, and the pain generally disappears when the injury heals. Chronic or persistent pain may range from mild to severe, and it is present to some degree for long periods of time. Some people with chronic pain that is controlled by medicine can have breakthrough pain. This occurs when moderate to severe pain "breaks through" or is felt for a short time. It may occur several times a day, even when the proper dose of medicine is given for chronic or persistent pain. Fortunately, pain can usually be controlled. Doctors, nurses, and all other members of the health care team are concerned with treating and controlling pain. Ongoing assessment of the types of pain that develop and change during the course of the cancer and its' treatment are essential to prescribing appropriate pain treatments. If pain is present, it can be caused by several factors, including those that have nothing to do with cancer. It is imperative that the cancer patient alerts the doctor immediately about any pain that they have. If cancer pain is left unattended, it can affect the patient's ability to work and participate in normal activities, as well as their quality of life. Not all people will be able to tolerate their drug treatment. Some people are allergic to certain medications. Some will eventually develop side effects from the medications. Some will tolerate one specific drug in a class of drugs, but not tolerate others in the same class. Some people may not be able to tolerate any of the drugs within a particular class. World Health Organization guidelines suggest that doctors try a particular drug in its' class to see if the patient will indeed tolerate it. The dosage is then increased until the patient gets either pain relief or intolerable side effects. Before abandoning that class of drugs entirely, another drug in that class will be tried. Sometimes the side effects can be managed with other treatments before discontinuing therapy.

Types of cancer related pain:

Nociceptive pain - also known as somatic pain, is caused by damage to tissue. It is often described as sharp, aching, or throbbing pain. Most of the pains we experience in everyday life – the pains from cuts and bruises, broken bones, or surgery – fall into this category of pain. It is often due to a cancer growing larger, cancer that has spread to the bones, muscles or joints, or a blockage of an organ or blood vessels. This type of pain responds well to most narcotic therapies.

Neuropathic pain – occurs when there is actual nerve damage. It may be caused by a tumor pressing on a nerve or a group of nerves, damage by chemotherapy drugs, surgery or radiation, or the direct invasion of cancers into the nervous tissue. Neuropathic pain is often unresponsive or resistant to narcotic therapies. Often times, this type of pain will require greater doses of narcotics than those used to control somatic pain. This pain is often described as a burning sensation, shooting, sharp, electrical, or lightening-like. When pain is caused by a tumor pressing on a nerve, radiation and/or chemotherapy may be used to shrink the size of the tumor. Surgery is then used to remove the tumor, hopefully removing the source of the problem and lessening the pain associated with it.

Incident pain – occurs when a patient moves or changes positions. Narcotics work well when the patient is lying down quietly, but the pain increases when they change positions. Treatment for incident pain is generally a nerve block which destroys the nerves causing the pain and is an invasive therapy that requires a pain specialist to administer it

Phantom pain – may occur if you have had an arm or a leg removed by surgery. You may still feel pain or other unusual or unpleasant sensations as if they were coming form the absent (phantom) limb. Doctors are not exactly sure why this occurs, but phantom limb pain is real; it is not "in your mind." This pain can also occur if you have had a breast removed – you may have a sensation in the site of the missing breast. No single pain relief method controls phantom pain in all patients all of the time. Many methods are used to treat this type of pain, including pain medicine, physical therapy, antidepressant medications, and transcutaneous nerve stimulation.

Pain is often made worse by intense worry and fear. Patients are overwhelmed by their own fear of dying, suffering, possible deformity, financial devastation, and isolation. It is not uncommon for a cancer patient to interpret new pain as the spreads of the disease or impending death. Support form family and friends is critical to help the patient to not feel alone and desperate. Non-drug treatments are now widely used to help manage cancer pain. There are many techniques that are used alone or with medicine. Some patients find that they are able to take a lower dose of medicine with such techniques.

The following are alternative methods for pain control:

- Exercise
- Relaxation / Meditation
- Imagery / Visualization

- Hypnosis
- Transcutaneous nerve stimulation
- Acupuncture / Acupressure
- Massage
- Talking with clergy or other spiritual advisors
- Music

Medicines Used to Relieve Pain

Nonopiods – are used for mild to moderate pain and include acetaminophen and nonsteroidal anti-inflammatory drugs (NSAID's), such as aspirin and ibuprofen. These can be bought over-the-counter (without a prescription). Acetaminophen relieves pain in a similar way to NSAID's, but does not reduce inflammation as well. Moderate amounts of alcohol can produce liver damage in people taking acetaminophen. Acetaminophen can also cover up a fever which may be a sign of infection and need to be treated. The most common side effect from NSAID's is stomach upset or indigestion, especially in older patients. NSAID's may also slow blood clotting, especially if you are undergoing chemotherapy. Certain conditions may be made worse by NSAID's or any product containing NSAID's. In general NSAID's should be avoided by people who:

- Are allergic to aspirin
- Are on chemotherapeutic drugs
- Are on steroid medicines
- Have stomach ulcers or history of ulcers, gout, or bleeding disorders
- Are taking prescription medicines for arthritis
- Are taking oral medications for diabetes or gout
- Have kidney problems

- Will have surgery within a week
- Are taking blood-thinning medication

Opiods - are used for moderate to severe pain and are also known as narcotics. Opiods are similar to natural substances (endorphins) produced by the body to control pain. Morphine, fentanyl, hydromorphone, oxydodone, and codeine are all included in this category of pain killers. Nonopiods are generally used in conjunction with these drugs to relieve pain.

Adjuvant medications – are not usually labeled as pain relievers, but do relieve pain. These include steroids, antidepressant, antihistamines, sedatives, and anticonvulsive medications.

Pain Medication Intolerance and Road Blocks

- Allergies
- Nausea
- Vomiting
- Constipation
- Sedation
- Difficulty in urinating
- Hallucinations
- Tolerance

Mental and Physical Fatigue

Fatigue, feeling tired and lacking energy, is the most common side effect reported by cancer patients. The exact cause is not always known. It can be due to the disease, chemotherapy, radiation, surgery, low blood counts, lack of sleep, pain, stress, poor appetite, along with many other factors. Fatigue from cancer feels different than fatigue of everyday life and levels of fatigue vary from one person to another. Patients with cancer have described it as a total lack of energy and have use words such as worn out, drained, and wiped out to describe their fatigue. It is not always relieved by rest and may last for months after treatment has ended. Severe fatigue gradually goes away as the tumor responds to treatment

Mental Fatigue – is noted by many patients with cancer in their inability to concentrate and focus. This is often referred to as "attentional" or "mental" fatigue. When you consider the many changes in ones' life that are associated with cancer, it's not hard to understand the demand for a great deal of mental exertion. It is estimated that cancer-related fatigue affects 76% of patients undergoing therapy. As if dealing with the cancer diagnosis is not enough to deal with in and of itself, these demands can place a patient at high risk for developing mental fatigue. It is necessary for the individual with cancer to focus on the reality of cancer and what it means to them and their loved ones. The focus can become overwhelming and the amount of energy it takes to deal with the emotions can be absolutely draining. Individuals are trying to process critical information about the disease at a time when emotions are already high. They are faced with a complete overhaul of their life as they know it. The demands may eventually exceed capacity. At this point, mental fatigue may result in loss of concentration.

It may become increasingly difficult to perform normal everyday tasks including self-care. This may affect a patients' quality of life due to their inability to perform normal activities that add meaning to their lives. For those with small children to take care of, this altered state of mind becomes even more frustrating. Patients often describe not being able to think clearly, forgetfulness, and difficulty in learning. The ensuing mental fatigue can lead the patient to feel that things are out of control. They become scared and feel helpless. They may become impatient and irritable causing a strain on their personal relationships.

Some ways to conserve mental energy include:

- Reduce time constraints and pressures
- Prioritize and save your energy for the most important things
- Break large tasks into smaller, easier to achieve goals
- Make reminder lists of things that must be done
- Keep things simple and free of distractions

- Take short naps or breaks, rather than one long rest period

Restoring mental energy is just as important as conserving it. To do so a patient should be able to choose activities that are interesting and will not allow them to get bored. They should be something enjoyable, other than the normal everyday routine. Activities like gardening, watching wildlife, watching fire in a fireplace, and playing with a pet can be very soothing. Exercise should be simple – walking in a natural environment, or slow, static stretching to soft music are good choices and should be encouraged 3-4 times per week for 20-30 minutes Patients should not be pushed to participate when they do not feel up to it.

Physical fatigue – During radiation treatment, the body uses a lot of energy healing itself. In addition to the mental fatigue from trips to receive treatment and stress related to the illness, the effects of treatment on normal cells all may contribute to physical fatigue. After a few weeks of radiation treatment most people will begin to feel tired. The level of tiredness will vary from one person to another. Generally, feelings of weakness will gradually go away after the treatment is finished. It is important during therapy not to try and do too much. Activities should be limited and leisure time should be for restful activities. Patients should not expect themselves to do what they usually would. More sleep at night and more rest during the day will be essential. The way patients adapt to their treatment varies from person to person. Some may continue to work a full-time job, while others will take several weeks off during treatment. Some people may choose to cut their hours back. But enjoy the camaraderie of their work environment. This is one time where the patient should not feel bad about asking family and friends to help with daily chores, shopping, housework, and even driving. Driving can be somewhat dangerous do to mental weariness that may accompany treatment.
Chemotherapy can reduce the bone marrow's ability to make red blood cells. Red blood cells transport oxygen to all parts of the body. If there are too few red blood cells, body tissues don't receive enough oxygen to do their job. This condition is known as anemia. Anemia can also be the result of an iron deficiency. Anemia is a common complication of cancer treatments. Although it is seldom life-threatening, anemia can make patients feel very weak and tired all of the time and has a severe impact on their quality of life. It is estimated that most patients with cancer will develop anemia at some point during the course of their disease and treatments. Additional symptoms include chills, dizziness, loss of appetite, inability to concentrate, chest pain, elevated heart rate and/or shortness of breath. Their doctor should be checking their blood cell count often during treatment. It is possible that if their red blood cell count falls too low, that they may need a blood transfusion. This is only done in severe, chronic cases. Transfusions can lead to additional complications such as fever, allergic disorders, infections, and suppression of the immune system. This is only seen in about 20% of the

cases. There are medications that stimulate red blood cell production that can be used as alternatives to a blood transfusion.

Some ways to conserve physical energy include:

- Get plenty of rest, including naps during the day.
- Limit daily activities to only the most important and necessary things.
- Ask family and friends to help out with things such as shopping, cooking, housework, and taking care of the children.
- Eat a healthy, well balanced diet. Seek the help of a nutritionist if necessary.
- When getting up from a sitting or lying position, move slowly. This will help prevent dizziness and the possibility of falling.
- Try easier or shorter versions of activities you enjoy.

Benefits of Exercise

Benefits of exercise in preventing cancer

In 1996, the first Surgeon General's report on physical activity and health was published, including the currently accepted public health recommendations for physical activity for general health, 20 minutes of moderate intensity activity – such as brisk walking – on most days of the week. This recommendation has been adopted by the American Cancer Society and is included in the current recommendations from the American Cancer Society in preventing cancer. Exercise has many proven health benefits for both preventing disease and promoting health and well being. There is substantial evidence that suggests that increasing physical activity, including structured exercise programs, is associated with lower rates of certain cancers. In particular, there is evidence that high levels of physical activity can work to prevent colon cancer. Cancers of the breast, prostate, lung, and uterus have also been linked to exercise-related prevention. In a large scale study of 17,148 Harvard alumni, men who burned as few as 500 calories a week in exercise – the equivalent of an hour's worth of brisk walking or less than ten minutes of waking a day – had death rates 15-20 percent lower than men who were almost completely sedentary. Men who burned 2,000 calories a week (about four hours of brisk walking per week) had about 35 percent lower cancer mortality. The researchers concluded that the more exercise you get, the lower your risk of premature death from cancer or heart disease. The Harvard study also found that the risk of colon cancer, the second leading cause of cancer-related death in the U.S., was dramatically reduced by exercise. Prostate cancer is the most common cancer affecting men today. In the Harvard study, alumni who expended greater than 4,000 calories per week (equivalent to about eight hours of brisk walking) were at a reduced risk of developing prostate cancer compared to their inactive counterparts. For women, a history of moderate, recreational exercise is associated with reduced risk of breast, uterine, cervical, and ovarian cancers, although not all studies have shown this effect. Findings from a 1993 study suggest that women engaged in moderate to high levels of physical activity may have a reduced risk of endometrial cancer. Currently, scientists are studying the biological impact that exercise has on the risk of cancer. Some of the methods that are being studied include:

- Maintenance of a healthy body weight and overall amounts of body fat.
- Maintenance of low levels of fat in and around the abdomen.
- Maintenance of the biological system that regulates blood sugar levels.
- Control of some tumor growth factors.
- Suppression of 'prostaglandins' (hormone-like substances that are released in greater quantities by tumor cells).

- Improved immune function, including increased levels of Natural Killer cells.
- Reduced symptoms of mild to moderate anxiety and depression (which may improve immune function and overall physiologic functioning).
- Increased levels of free radical scavengers to assist the body in preventing DNA damage

It is not clear exactly how high amounts of physical activity work to prevent cancer. We know that exercise can help prevent obesity, which is related to some types of cancers. It can also change the body's hormone levels, which might also have a favorable effect. Exercise, by speeding up metabolism, is generally believed to speed up the passage of ingested foods through the colon – thus reducing the amount of time the colon mucosal lining is in contact with possible carcinogens. Additionally, those who engage in a high level of physical activity are much less likely to smoke cigarettes, the single largest contributor to cancer.

Benefits of exercise during treatment

Starting or maintaining an exercise program after cancer diagnosis results in patients who are stronger both mentally and physically, concludes a statistical analysis of 24 studies. Kerry Courneya of the University of Alberta, Canada led the research, which is published in the Annals of Behavioral Medicine. Courneya says "Cancer diagnosis and its' treatments are often associated with negative side effects that diminish the quality of life. Overall, studies have consistently demonstrated that physical exercise following cancer diagnosis has a positive effect on the quality of life." The various studies mention increased stamina, increased functional capacity, strength, self-esteem, improved treatment tolerance, and satisfaction with life, and decreased pain. Psychological changes, including a decrease in total mood disturbances, decrease in depression, and fewer problems sleeping were noted between the exercise and non-exercise groups. It has also been noted that increased physical activity has been associated with less fatigue during and after chemotherapy and radiation. The specific exercise "dose" (frequency, intensity, and duration of sessions) needed to improve physical and psychological functioning in cancer patients probably differs according to specific treatment, cancer type, and individual response to treatment. Some forms of cancer treatment, particularly those that are used to treat childhood cancers, have been found to have long-term negative effects on the heart and lungs. This makes it even more important to exercise regularly, but it may important to do so under medical supervision.

Benefits of exercise during recovery from surgery

After cancer surgery exercise plays an invaluable role in helping one return to the strength and fitness level that was maintained prior to surgery. In many cases, due to lack of physical activity prior to surgery, patients are able to reach new heights in strength, flexibility, and cardiovascular conditioning. There are certain postural implications that often arise after mastectomy and lymph node dissection that are often compounded by reconstruction and radiation. After years of working with cancer survivors, we declare with certainty that most of these issues can be dramatically improved upon if not entirely corrected, through the proper combination of stretching and strengthening. Anytime there is an amputation, it will ultimately result in some type of muscle imbalance. These issues will not correct themselves. Unfortunately, even patients who undergo physical therapy are released long before they are fully recovered, leaving the patient to go it alone in determining how to resume normal activities. In addition, when patients receive radiation to a particular area, there is bound to be some tightness, perhaps even scar tissue, where they received treatment. This can cause tightening in that area, and depending on where it is, can also contribute to many postural deviations. These postural imbalances are notable in most people due to everyday circumstances i.e.; working at a computer all day, holding a phone between your ear and your shoulder, sitting at a desk all day, holding a baby on one hip etc… Not only are they compounded by the surgery and radiation, but they can create a chain reaction, leading to neck, back, hip, knee, and even ankle pain. A thorough postural assessment can determine what areas need to be stretched to relieve tightness and spasm and which need to be strengthened to create a counter balance. Let's not forget about the many benefits of cardiovascular conditioning. Many of your clients may still be suffering from fatigue long after their treatment has ended. Cardiovascular training, biking, walking, running, etc., will produce endorphins that will give them much needed energy. Unfortunately chemotherapy and radiation can have a detrimental effect on the heart and lungs. The good news is that both can be strengthened through a regular cardiovascular exercise program. Swimming can provide an excellent source of relief for tight muscles without putting excessive strain on them. The buoyancy of the water allows for a wonderful workout that allows clients to focus on range of motion for their arms and shoulders. This is highly recommended for breast cancer patients, particularly those who have undergone an axillary node dissection. Swimming should not be limited only to breast cancer patients, however, for it has benefits for everyone. Those clients suffering from arthritis will want to make sure the water is at least eighty degrees.

Cancer Treatment & Weight Management

Weight gain after a cancer diagnosis

In cancer survivors, weight gain may lead to the development of other diseases as well as lower cancer-related survival and overall survival. Additionally, cancer survivors are at greater risk for developing second cancers as well as other diseases, such as heart disease and diabetes. It is well documented that heart disease and diabetes are clearly linked to weight gain.

Certain cancer treatments may alter the patients' body composition. Studies show that people with breast cancer, prostate cancer, non-small cell lung cancer, and acute lymphoblastic leukemia, who undergo chemotherapy, hormone therapy, or radiation therapy to the head, show unfavorable changes in their body composition. Their body fat increases and their lean muscle decreases. While their weight may not fluctuate dramatically on the scale, because of the increase in body fat, it is very likely that they will go up a size or two in their clothing.

The most important factors for weight gain in cancer patients seem to be the decrease in physical activity and the resulting lower basal metabolic rate. For those that are able to tolerate food, and have a propensity for over-eating, this will be a factor as well. In order to maintain their current body weight, patients should reduce their normal calorie intake and/or increase exercise, such as resistance training, that helps build muscle. Unfortunately, depending on the type of treatment, as well as the age and lifestyle of the individual, options may be somewhat limited. This is where the Cancer Exercise Specialist can make recommendations for the frequency, intensity, and duration of the exercises; as well as which are best suited for the individual client.

General recommendations for weight control

Cancer patients are not alone in the never-ending struggle to reach and maintain their ideal body weight. It becomes even more of a priority, however, because a health body composition is associated with cancer prevention, more effective cancer treatment, the prevention of type II diabetes and heart disease, improved overall health and survival, increased ability to perform activities of daily living, and better quality of life.

It is easy to gain weight in a society that boasts "super-size" portions, "all you can eat buffets," and a "more is better" mentality. Making good choices required discipline and self-control, but also being an educated consumer.

Here are some tips for making better choices:

- Choose foods with lower calorie content, such as vegetables, fruits, whole grains, and soups. Some of these foods also help a person feel "full" faster due to their high fiber content.
- Limit foods and beverages that are high in sugar and fat. Drink plenty of water (which will also produce the feeling of fullness without the added calories)
- Balance the calories from foods and beverages with the amount of calories burned through physical activity. 3500 calories equals one pound. Therefore, eating a piece of cake that is 600 calories will require 600 calories of physical activity in order to burn it off. If the daily caloric intake exceeds the daily caloric expenditure, weight gain will occur.
- Increase levels of physical activity. General recommendations are to aim for 30 to 60 minutes per day of moderate-to-intense exercise 5 or 6 times a week. Patients that are suffering from extreme fatigue and other treatment side-effects should try to do some form of physical activity everyday; even if it's just a ten minute walk. For patients who were sedentary prior to their diagnosis, even a slight increase in physical activity will show marked improvement. On the other hand, those who were very active prior to treatment will probably be frustrated with their inability to perform at the level they have become accustomed to. It will be imperative for them to reduce their caloric intake in accordance with the reduction in physical activity in order to maintain their weight.
- If your client was overweight or obese prior to their cancer diagnosis, they should take steps to lose weight through nutrition and exercise. Body composition is the key component. If the scale does not show a marked change, but their lean muscle to body fat ratio improves, that is still a step in the right direction.

Many people with cancer find themselves needing to gain weight. As a result of certain cancer surgeries and treatments, patients may have loss of appetite. This is exacerbated by nausea, vomiting, mouth sores, difficulty swallowing, and loss of taste. A registered dietitian who specializes in working with cancer patients can make recommendations for adding calories and improving nutrition. Counseling provided by a registered dietitian can help patients and survivors who have completed treatment lose or gain weight. Weight loss plans, that are initiated prior to treatment and include exercise, have also helped people with cancer avoid weight gain during chemotherapy. Weight loss in people with cancer or cancer survivors should be closely monitored and reported to their doctor, and, like the general population, no more than 2 pounds should be lost each week.

Conquering Cancer with Nutrition

Conquering Cancer With Nutrition

Glenn B. Gero, N.D., R.N.C., M.H., C.E.S., C.L.C.

There's no longer a question that a good diet is essential for optimal health. We know that the foods we eat and our lifestyle affect our health. Food, nutrition and physical activity are crucial to our general health and well-being. The same way that our lifestyle choices affect our daily health, it also affects our long-term risk of developing diseases such as cancer. There have been scores of scientific research validating the association of poor dietary choices and the onset of degenerative diseases including cancer. As a matter of fact, according to The American Institute for Cancer Research, it is estimated that between 60 and 70 percent of all cancers have been directly linked to our daily dietary and lifestyle habits. On the other side of the coin, according to the Institute, we can achieve dramatic reductions in our cancer risk by making small adjustments to our daily dietary choices. A significant body of evidence has demonstrated that certain foods, for example, can offer benefits to people already afflicted with cancer either because they help to treat the condition, bolster immune response or because they can potentiate the effectiveness of other conventional or integrative therapies.

Primary Nutritional Goals

There are three main nutritional goals for someone living with cancer. They are 1) to maintain a healthy bodyweight; 2) to select a nutritional plan that will supply the body with fuel and nutrients for repair and healing and aid in the body's ability to eliminate toxins; and 3) to prevent recurrence of the cancer and the development of the second malignancy. While it may be overwhelming to administer radical alterations in food choices, it's got to be understood that if one refuses to make substantive modifications in their lifestyle, everything will stay the same; hence, one's cancer risk and the potential for recurrence will remain unaltered.

American Adults are Fat

In fact, 64 percent of us are either overweight or obese. The number of obese people in this country has nearly doubled since 1980. Unless we learn to eat less and exercise more, the consequences will be grave. Science has clearly drawn an association between obesity and cancer risk. A recent report by the World Health Organization's International Agency for Research on Cancer estimates that being overweight and inactive accounts for one-fifth to one-third of all breast, colon, endometrial, kidney and esophageal cancers. There is also strong evidence of the association of obesity and cancers of the pancreas, uterus, prostate, ovary and liver. In the U.S. alone, that represents between 102,000 and 135,000 cases each year.

Researchers predict that at the current rise in the incidence of obesity, cancer rates may escalate by 50 percent by the year 2020. The premise is that being significantly overweight and inactive produces dramatic hormonal and metabolic changes that create the optimum environment for the onset and proliferation of cancer cells. Simply put, the secret to effective and sustained weight control is a five-part process.

1. Move toward a plant-based diet - consuming mineral rich foods fulfill the body's need for nutrients and provides lower calorie and higher fiber foods that usually reduce unhealthy cravings.

2. Watch your portion sizes – condition yourself to eat only when you're hungry and only as much as your body requires. Pausing after each forkful to take a couple of deep breaths will enhance portion consciousness.

3. Keep physically active – exercise burns calories, regulates metabolism, relieves stress and may reduce the urge to eat.

4. Go slowly – crash diets, skipping meals and excessive exercise are usually short-lived and fail 97 percent of the time. A carefully planned program of a gradual weight reduction, focusing on healthy eating and exercising will provide sustained results.

5. Learning to control our response to life's stresses may play a major role in weight maintenance and uncontrolled cravings and binging. It is the regrets of the past and the fears of the future that proliferate much of our unresolved stress. Learning to live in the present can often have a dramatic effect on our ability to reduce stress in our lives.

Achieving and maintaining a healthy body weight is an essential component of cancer prevention and treatment. Gradually reducing body fat and keeping it off is a strong step in the direction to help protect against cancer and assist in its treatment.

Nutritional Cancer Fighting Tools

Although every cancer patient requires an individual nutritional and therapeutic protocol, there are some important general guidelines which may give the body the best chance of preventing or recovering from cancer or other debilitating degenerative diseases.

1. Eat your greens, reds, oranges, yellows and purples

A diet rich in fruits and vegetables is the best bet for preventing cancer. That fact has been supported and endorsed by the U.S. government agencies and by virtually every major medical organization, including the American Cancer Society. Selecting foods of different colors – red, orange, yellow, green and purple – you'll be getting a full spectrum of compounds that contribute to optimal health and have the greatest potential to protect and possibly even reverse catastrophic diseases like cancer.
The American Institute for Cancer Research reports, "if the only change people made was to eat at least five servings of fruits and vegetables each day, cancer rates could drop by at least 20 percent." Eating fruit and vegetables are our most natural and absorbable source of vitamins and minerals. This should be our primary sources of these nutrients. Synthetic pills that are produced in a laboratory cannot replace the vitality that real food can offer. These plant-based whole foods containing natural substances are called phytochemicals.

These phytochemicals have demonstrated the potential to modulate cancer development. There are many biologically plausible reasons why consumption of plant foods might slow or prevent the appearance of cancer. These include the presence in plant foods of such potentially anticarcinogenic substances as carotenoids, chlorophyll, flavonoids, indole, isothiocyanate, polyphenolic compounds, protease inhibitors, sulfides, and terpens. Many experts refer to these compounds as chemopreventers to emphasize their potent anti-cancer effects. While these powerful phytochemicals work in harmony with antioxidants like vitamin C, vitamin E and selenium, they offer even considerably greater protection against cancer.

The specific mechanisms of action of most phytochemicals in cancer prevention are not yet clear, but they appear to be varied. Considering the large number and variety of dietary phytochemicals, their interactive effects on cancer risk may be extremely difficult to assess. Phytochemicals can inhibit carcinogenesis by inhibiting certain cancer causing enzymes, curtail the mutation of DNA, suppress the abnormal proliferation of early cancerous lesions, and inhibit certain properties of the cancer cell.

Optimally, we should consume at least ten servings of fruits and vegetables each day. Choose a variety of colors of produce to get the most benefit. An orange with breakfast, apple for mid-morning snack, tossed salad and mixed fruit for lunch, raw veggies for an afternoon snack, two types of vegetables for supper and a late-night banana add up to ten.

2. Consume the fats that heal

There is indisputable evidence that a diet rich in saturated fats and cholesterol have been linked to cancer. Both the American Cancer Society and the National Cancer Institute recommend a diet that supplies less than 30 percent of dietary calories from fat. Just as important, however, is the type of fat consumed. Along with decreasing total fat intake, it is important to increase the intake of omega-3 fatty acids.

Cancer scientists continue to find cancer-fighting potential in omega-3 fatty acids. Omega-3s are the type of polyunsaturated fat found mainly in fatty fish like salmon and, to a lesser extent, in certain vegetables, nuts, seeds and oils. Human and laboratory studies show evidence that omega-3s may lower cancer risk.

W. Elaine Hardman, Ph.D., a researcher at the Pennington Biomedical Research Center of Louisiana State University explains, "populations in countries that consume high amounts of omega-3 fatty acids from fish have lower incidences of breast, prostate and colon cancer than in countries than consume less omega-3s." In laboratory studies funded by the American Institute for Cancer Research, Hardman has found that supplementing the diet with omega-3s can reduce occurrence of tumors.

Omega-3s may also help cancer therapy's effectiveness. In other laboratory studies, Dr. Hardman found that adding fish oil to the diet can slow tumor growth, help chemotherapy drugs work more effectively and reduce side effects from other cancer treatments.

Unlike fish oil which is high in omega-3 polyunsaturated fatty acids, fats that are high in omega-6 polyunsaturated fatty acids (like corn oil), can proliferate the growth of tumors. Using a chemical carcinogen-induced cancer model, researchers found that a high intake of fish oil significantly lowered the cancer incidence in animal studies as compared to animals fed either low fat diets or diets high in corn oil. By implanting human tumors into immune-deficient mice, researchers have found that a high fish oil diet can slow tumor growth. These results suggest that fish oil can be used for both prevention and treatment of cancer.
Although there is no clear mechanism to explain fish oil's significant anticancer effects, researchers have uncovered several potential models of action:

- Alteration of cell membrane composition. After ingestion, fish oil is easily incorporated into cell membranes (especially tumor cells) which changes the cell membrane composition. This alteration will change the cell's response to growth factor, hormones, antibodies, etc.

- Inhibition of prostaglandin production. Prostaglandins can stimulate tumor cell growth. Fish oil can inhibit the enzyme responsible for prostaglandin synthesis called prostaglandin synthase. After a high intake of fish oil, prostaglandin (especially in the tumor cells) is decreased significantly, which in turn slows tumor growth.

- Fish oil can enhance immune system stimulation.
- Hormone profile changes, which may provide important benefits for hormone-related cancers like breast cancer.
- Tumor cell toxicity, probably by causing lipid peroxidation (or oxidative deterioration) in the tumor cells.

One of the big concerns in cancer treatment is metastasis, the process by which tumor cells spread from the primary location to distant parts of the body. Metastasis is increased by a high intake of omega-6 fatty acids (e.g., corn oil), but is inhibited by fish oil. Using an immune deficient mouse implanted with human breast cancer, researchers found that feeding a high fish oil diet (23%) to the mice significantly reduced human breast cancer cell metastasis to the regional lymph nodes and lungs. This indicates the significant beneficial effects of fish oil supplementation in cancer treatment.

Researchers at Allie M. Lee Cancer Research Laboratory at the University of Nevada, Reno, first declared that fish oil supplementation may be of benefit in cancer chemotherapy. By using a human breast cancer model, they found that feeding the animals a high fish oil diet both slowed the tumor growth and increased the tumor responsiveness to chemotherapy drugs by altering the drug activating systems. They also found that a high fish oil diet can significantly protect the host animals against the toxicity of chemotherapy drugs.

3. Reduce the exposure to pesticides

In the United States, more than 1.2 billion pounds of pesticides and herbicides are sprayed or added to food crops every year. Exposure to these chemicals damages the body's detoxification mechanisms, thereby increasing the risk of getting cancer and other serious diseases.

Let's think about the seriousness of this situation. Farmers in this country live a fairly healthy diet compared to those living in metropolitan areas. They have constant access to fresh fruit and vegetables, they breathe clean air, get plenty of exercise, they have a lower rate of cigarette smoking and significantly less alcohol and drug dependency. Yet studies indicate that farmers have a higher risk of developing leukemias, lymphomas and cancers of the stomach, prostate, brain and skin.

There is also significant evidence that there is a correlation between exposure to pesticides and the risk of non-Hodgkin's lymphoma. This blood cancer currently accounts for about 3 percent of all cancers diagnosed in the United States.

While pesticides may increase risk, they are not necessarily the only factor involved. The presence of pesticides should not deter us from eating fruit and vegetables. As a matter of fact, the levels of pesticides are lower in these foods than in those found in animal fats, meat, cheese, whole milk and eggs. Additionally, the antioxidants found in fruits and vegetables are necessary to help the body deal with the pesticides.

Here are some recommendations to reduce exposure of pesticides:

- Don't over consume animal fats, non-organic eggs and conventional dairy products.
- Buy organic products when possible.
- Peel skin from outer layers (which also contain many nutrients) or wash produce thoroughly with a biodegradable cleanser, then rinse.

4. Avoid foods that create a "nutritional debt"

Imagine that you just got a new job and, after completing your first 40-hour work week, your boss approaches you with some bad news. "The company isn't doing very well and we, unfortunately, don't have enough money to pay you." This is analogous to our refined, high calorie, low nutrient junk food diet. Our body must expend energy to digest food that offers little in return, robbing us of the vitamins, minerals, phytonutrients and enzymes that we need to perform all of our physiological functions optimally.

Beware of foods high in sugars, refined flours, excess sodium, artificial sweeteners, trans or hydrogenated fats, corn syrups and deep fried anything. Reach instead for the "good stuff." A diet high in nutrient dense foods will protect immune response, heighten energy production, enhance mood and balance blood sugar and insulin levels, all of which may reduce our risk of developing cancer.

Some guidelines for making better nutritional choices include:

- Read labels. If sugar, partially-hydrogenated fat, salt or "enriched flour" is listed as one of the first several ingredients, don't buy it. If your foods are naturally rich in nutrients, they don't need to be enriched, salted, sweetened or hydrogenated. Look for grain products that contain 3 grams of fiber or more per serving.

- Be aware of ingredients such as sucrose, glucose, maltose, lactose, fructose, corn syrup or white grape juice concentrate, which indicates that sugars have been added.

- Look for the percentage of fat calories to total calories, as well as the number of fats grams per serving. For every 5 grams of fat in a serving, you are eating the equivalent of one teaspoon of fat.

5. Reduce animal products

One basic truth that seems to be confirmed from study to study is the fact the higher the intake of meat and other animal foods, the higher the cancer risk, especially for the major cancers, such as colon, breast, prostate and lung cancers.

According to noted naturopath Dr. Michael Murray, there are many reasons for this association. "Meat," Dr. Murray explains, "lacks the antioxidant and phytochemicals that protect us from cancer. At the same time it contains lots of saturated fat and other potentially carcinogenic compounds—including pesticide residues, heterocyclic amines and polycyclic aromatic hydrocarbons, which form when meat is grilled, fried or broiled. The more well done the meat, the higher the level of amines."

While there is significant controversy, the actual risks associated with a diet high in animal products is associated with our demand for tender cuts of meat, which have compromised the nutritional value and safety of our animal food supply. Grain-fed cattle are tortured and restrained in tiny cubicles, injected with hormones and fed an unnatural grain and dairy diet. These measures have altered the fat composition of domestic cattle. Domestic beef contains primarily saturated fats and virtually no beneficial omega-3 fatty acids, while the fat of range-fed or wild animals contain more than five times the polyunsaturated fat per gram and has substantial amounts (about 4 percent) of omega-3 fatty acids.

Some sensible guidelines in choosing animal products include:

- Limit daily portions to about the size of the palm of your hand.

- Avoid overcooked, charbroiled, deep-fried or overly-fatty meats.

- Read labels and avoid meats preserved with nitrates, nitrites or msg.

- Buy organic grass-fed, free-range poultry, beef, buffalo, ostrich or venison.
- Only consume organic eggs from free-range chickens.

6. *Select foods and spices that will help detoxify the body*

The National Cancer Institute estimates that roughly one-third of all cancer deaths may be diet related. What you eat can hurt you, but it can also help you. Many of the common foods found in grocery stores or organic markets contain cancer-fighting properties, from the antioxidants that neutralize the damage caused by free radicals to the powerful phytochemicals that scientists are just beginning to explore. There isn't a single element in a particular food that does all the work. The best thing to do is eat a variety of foods.

The following foods have the ability to detoxify the body, help stave off cancer and some can even help inhibit cancer cell growth or reduce tumor size.

Avocados are rich in glutathione, a powerful antioxidant that attacks free radicals in the body by blocking intestinal absorption of certain fats. They also supply even more potassium than bananas and are a strong source of beta-carotene. Scientists also believe that avocados may also be useful in treating viral hepatitis (a cause of liver cancer), as well as other sources of liver damage.

Broccoli, cabbage and cauliflower have a chemical component called indole-3-carbinol that can combat breast cancer by converting a cancer-promoting estrogen into a more protective variety. Broccoli, especially sprouts, also have the phytochemical sulforaphane, a product of glucoraphanin - believed to aid in preventing some types of cancer, like colon and rectal cancer. Sulforaphane induces the production of certain enzymes that can deactivate free radicals and carcinogens. The enzymes have been shown to inhibit the growth of tumors in laboratory animals. However, be aware that the Agriculture Department studied 71 types of broccoli plants and found a 30-fold difference in the amounts of glucoraphanin. It appears that the more bitter the broccoli, the more glucoraphanin it contains. Broccoli sprouts have been developed under the trade name BroccoSprouts that have a consistent level of sulforaphane - as much as 20 times higher than the levels found in mature heads of broccoli.

Carrots contain a lot of beta carotene, which may help reduce a wide range of cancers including lung, mouth, throat, stomach, intestine, bladder, prostate and breast. A substance called falcarinol that is found in carrots has been found to reduce the risk of cancer, according to researchers at Danish Institute of Agricultural Sciences (DIAS). Kirsten Brandt, head of the research department, explained that isolated cancer cells grow more slowly when exposed to falcarinol. This substance is a polyacethylen, however, so it is important not to cook the carrots.

Cayenne (Red Pepper) contains a chemical, capsaicin, which may neutralize certain cancer-causing substances (nitrosamines) and may help prevent cancers such as stomach cancer.

Figs apparently have a derivative of benzaldehyde. It has been reported that investigators at the Institute of Physical and Chemical Research in Tokyo say benzaldehyde is highly effective at shrinking tumors. Fig juice, additionally, is also a potent bacteria killer in test tube studies.

Flax contains lignans, which may have an antioxidant effect and block or suppress cancerous changes. Flax is also high in omega-3 fatty acids, which are thought to protect against colon cancer and heart disease.

Garlic has immune-enhancing allium compounds (dialyl sultides) that appear to increase the activity of immune cells that fight cancer and indirectly help break down cancer causing substances. These substances also help block carcinogens from entering cells and slow tumor development. Diallyl sulfide, a component of garlic oil, has also been shown to render carcinogens in the liver inactive. Studies have linked garlic, as well as onions, leeks and chives to a lower risk of stomach and colon cancer. According to a report in the October 2000 issue of the *American Journal of Clinical Nutrition*, people who consume raw or cooked garlic regularly face about half the risk of stomach cancer and two-thirds the risk of colorectal cancer as people who eat little or none. Their studies didn't show garlic supplements had the same effect. It is believed garlic may help prevent stomach cancer because it has anti-bacterial effects against a bacterium, helicobacter pylori, found in the stomach and known to promote cancer there.

Grapefruits, like oranges and other citrus fruits, contain monoterpenes, believed to help prevent cancer by sweeping carcinogens out of the body. Some studies show that grapefruit may inhibit the proliferation of breast-cancer cells in vitro.

Grapes, red contain bioflavonoids, powerful antioxidants that work as cancer preventives. Grapes are also a rich source of resveratrol, which inhibits the enzymes that can stimulate cancer cell growth and suppress immune response. They also contain ellagic acid, a compound that blocks enzymes that are necessary for cancer cells - this appears to help slow the growth of tumors.

Kale has indoles, nitrogen compounds which may help stop the conversion of certain lesions to cancerous cells in estrogen-sensitive tissues. In addition, isothiocyanates, phytochemicals found in kale, are thought to suppress tumor growth and block cancer-causing substances from reaching their targets.

Licorice root has a chemical, glycyrrhizin, that blocks a component of testosterone and, therefore, may help prevent the growth of prostate cancer. Excessive quantities of licorice, however, may cause high blood pressure.

Mushrooms - There are a number of mushrooms that appear to help the body fight cancer and build the immune system - Shiitake, maitake, reishi, Agaricus blazei Murill, and Coriolus Versicolor. These mushrooms contain polysaccharides, especially lentinan, powerful compounds that help in building immunity. They are a source of Beta Glucan. They also have a protein called lectin, which attacks cancerous cells and prevents them from multiplying. They also contain Thioproline. These mushrooms can stimulate the production of interferon in the body.

Nuts contain the antioxidants quercetin and campferol that may suppress the growth of cancers. Brazil nuts contain 80 micrograms of selenium, which is important for those with prostate cancer.

Oregano, as reported in the *Journal of Agriculture and Food Chemistry* has the potential to destroy human cancer cells. Additionally, oregano has powerful antioxidant properties, due in large part to its rosmarinic acid (RA) content. Oregano oil increases oxygen levels in the blood, giving more energy, better athletic performance and mental clarity.

Papayas have vitamin C that works as an antioxidant and may also reduce absorption of cancer-causing nitrosamines from the soil or processed foods. Papaya contains folacin (also known as folic acid), which has been shown to minimize cervical dysplasia and certain cancers.

Raspberries contain many vitamins, minerals, plant compounds and antioxidants known as anthocyanins that may protect against cancer. According to a recent research study reported by *Cancer Research* 2001;61:6112-6119, rats fed diets of 5% to 10% black raspberries saw the number of esophageal tumors decrease by 43% to 62%. A diet containing 5% black raspberries was more effective than a diet containing 10% black raspberries. Research reported in the journal *Nutrition and Cancer* in May 2002 shows black raspberries may also thwart colon cancer. Black raspberries are rich in antioxidants, thought to have even more cancer-preventing properties than blueberries and strawberries.

Rosemary may help increase the activity of detoxification enzymes. An extract of rosemary, termed carnosol, has inhibited the development of both breast and skin tumors in animals. We haven't found any studies done on humans. Rosemary can be used as a seasoning. It can also be consumed as a tea: Use 1 tsp. dried leaves per cup of hot water; steep for 15 minutes.

Seaweed and other sea vegetables contain beta-carotene, protein, vitamin B12, fiber and chlorophyll, as well as chlorophylones - important fatty acids that may help in the fight against breast cancer. Many sea vegetables also have high concentrations of the minerals potassium, calcium, magnesium, iron and iodine.

Sweet potatoes contain many anticancer properties, including beta-carotene, which may protect DNA in the cell nucleus from cancer-causing chemicals outside the nuclear membrane.

Teas: Green Tea and Black tea contain certain antioxidants known as polyphenols (catechins) which appear to prevent cancer cells from dividing. Green tea is best, followed by our more common black tea (herbal teas do not show this benefit). According to a report in the July 2001 issue of the *Journal of Cellular Biochemistry*, these polyphenols that are abundant in green tea, red wine and olive oil, may protect against various types of cancer. Dry green tea leaves, which are about 40% polyphenols by weight, may also reduce the risk of cancer of the stomach, lung, colon, rectum, liver and pancreas, study findings have suggested.

Tomatoes contain lycopene, an antioxidant that attacks roaming oxygen molecules, known as free radicals, that are suspected of triggering cancer. It appears that the hotter the weather, the more lycopene tomatoes produce. They also have vitamin C, an antioxidant which can prevent cellular damage that leads to cancer. Watermelons, carrots and red peppers also contain these substances, but in lesser quantities. It is concentrated by cooking tomatoes (tomato paste being the richest source). Scientists in Israel have shown that lycopene can kill mouth cancer cells. An increased intake of lycopene has already been linked to a reduced risk of breast, prostate, pancreas and colorectal cancer. (Note: Recent studies indicate that for proper absorption, the body also needs some oil along with lycopene.)

Tumeric (curcuma longa), a member of the ginger family, is believed to have medicinal properties because it inhibits production of the inflammation-related enzyme cyclo-oxygenase 2 (COX-2), levels of which are abnormally high in certain inflammatory diseases and cancers, especially bowel and colon cancer. In fact, a pharmaceutical company, Phytopharm, in the UK, hopes to introduce a natural product, P54, that contains certain volatile oils, which greatly increase the potency of the turmeric spice.

Turnips are said to contain glucose molaes which is a cancer fighting compound.

Water is essential for life. The average amount of water in the body is about 10 gallons. We need to drink at least 48 ounces of water per day to replace the water that is lost through urination, sweat and breathing. Pure water is a universal detoxifier. In the body, it helps to remove waste materials and clear out toxins. It also helps carry oxygen and nutrients to all cells. Maintaining proper body hydration provides a foundation for both health and wellness.

Supplemental Insurance for Preventing Cancer

Dietary supplements are secondary to a sound diet plan because we can't guarantee that even the best diet offers the optimum nutrients our bodies need. Depleted soils, environmental toxins, impure water, physical demands and psychological stresses heighten our nutritional requirements. The following few supplements are some of the key recommendations which may significantly reduce the risk of cancer:

A food-based, multiple vitamin and mineral, especially one that offers adequate levels of anti-oxidants, serves as insurance against major nutrient deficiencies. Being food-based, they generally include important phytonutrients and detoxifying enzymes. Additionally, having a natural food base enhances its absorbability.

Green drinks are made by drying the juice from one or more of several plants. As these plants are typically 90% water, the usual serving (1/2 ounce or 20 capsules) of the green powder roughly equals one serving of a green vegetable. An extra serving of green vegetable each day doesn't sound like much, but on average, we Americans eat only two servings per day. As mentioned above, according to the National Cancer Institute, even a small increase in vegetable intake would reduce the risk of ovarian cancer, for instance, by about 20%.

Chlorophyll is the substance all green drinks have in common. Chlorophyll structure is almost identical to hemoglobin, except chlorophyll has a magnesium atom where hemoglobin has an iron atom. Our bodies require magnesium to utilize the energy we obtain from our food, and chlorophyll from plants is an important dietary source of this essential mineral.

The plants most often used to make green drinks are two single-cell organisms, chlorella and spirulina, and the young leaves of alfalfa, barley or wheat. Besides being rich sources of magnesium, these plants supply reasonable amounts of protein, vitamin E, and essential fatty acids such as gamma linolenic acid, an essential fatty acid that is hard to obtain in the typical American diet.

Just as it is prudent to consume a varied diet, it is sensible to use a mixture of the dried greens. A wide variety of other foods are often mixed in with several of green foods, which is okay as long as the greens are the main ingredient

In addition to known nutrients, chlorella also has something called Chlorella Growth Factor that may aid in human tissue repair. Researchers have also described complex sugars in chlorella that are 100 and 1,000 times more powerful than those currently used clinically for cancer immunotherapy. These complex sugars appear to be cell-to-cell messenger chemicals that stimulate white cells.

It is less expensive and tastier, no doubt, to acquire nutrients from whole foods, but if your lifestyle makes it prohibitively difficult to eat as you should, we are fortunate that in America another option exists.

Probiotics or beneficial bacteria are capable of altering certain enzymes that turn procarcinogens into carcinogenic agents. The "bad" bacteria that secrete these destructive enzymes include clostridium and certain bacteroides, among others. Obviously, the more dangerous enzymes that are present in our gastrointestinal tract, the greater our risk harboring cancer-causing substances. The ability of these active super strains of beneficial bacteria that can neutralize these harmful enzymes is one of the most important contributions to cancer prevention. The immune system's workload is further complicated by the need to cleanse the body of an increasing number of extraneous pollutants and contaminants found in the environment and the food chain. Overloaded with work, the immune system needs all the help it can get from our friendly bacteria. When disease-causing aliens are able to permeate the intestinal walls and enter the bloodstream, the immune system must spring into action. As long as strong colonies of friendly bacteria line the intestinal tract in full force, these harmful micro-organisms will not be able to get through, thus lightening the already heavy workload of the immune system.

Through scientific research, it has been determined that friendly bacteria can reduce the threat of potential cancer-causing agents in the body and increase the body's immune system to transform these agents into inactive carcinogens. Some cancer risk factors are under our control, especially diet. Supplementing the diet with probiotics, in addition to healthy dietary choices, is one way to help lower our risk of getting cancer. Additional supplements should be recommended, by your healthcare practitioner, on a case-by-case basis.

The information contained in this chapter is for educational purposes only and should not be deemed medical advice. These ideas are not intended to replace the advice of a qualified healthcare practitioner. Neither the authors of this book or the publisher shall be liable or responsible for any loss, injury or damage allegedly arising from any information or suggestions in this book.

Working With the Medical Community

As Cancer Exercise Specialists, we want to become the next step in the health care continuum. This means that we need to gain the respect of the health care providers so they feel comfortable referring their patients to us for care. The continuum will typically follow this order:

- Doctors
- Physical Therapists, Occupational Therapists, Nurses
- Cancer Exercise Specialist

There are many ways in which you can gain free advertising and publicity.

- Breakfast/lunch meetings
    - Call the office manager to set up a time that you can meet with the doctor and/or staff to introduce yourself and your program – tell them you'll bring coffee & bagels or sandwiches (based on the time of day you're meeting)
    - Prepare for your meeting; figure out a ten-minute effective presentation. Bring brochures and business cards that outline your program and its' benefits.
    - Make sure you know what you are talking about before you take this step!
- Support groups
    - Get online and look up cancer support groups in your area
    - Call the director and offer to come in and give a complimentary 10-15 presentation on exercise programming for that particular type of cancer (there are support groups for all typed of cancer)
    - Prepare your presentation and bring brochures and business cards that outline your program and its' benefits
- Online directory
    - When you pass the CES examination, you will be added to our internet directory
    - Make sure to update us is your contact information changes so we can change it on the website
    - This is a free resource to advertise you as a CES and will lead to potential referrals for clients

- Media – newspaper, television, and radio
    - Contact all of the local radio and television stations as well as the local newspapers. Magazines are a bit more difficult, but certainly worth a try.
    - Let them know about who you are, your extensive training in the area of cancer and exercise, and that you are one of the few (perhaps the first) in your area to provide such services.
    - Re-create the press release in this book to be representative of you and your program. Send it out to all local radio and television stations as well as the local newspapers
- Always have your client complete the appropriate paperwork and make sure that you complete a thorough assessment of your client as well as keeping daily notes as to any problems, aches, pains, or progress that you may be aware of.
    - Health History
    - Liability Release
    - Medical Clearance Form
- Follow up by keeping Doc's in the "loop." Make them feel like they are still part of the team and that you value their opinion when it comes to their patients' health.
    - Initial assessment and introduction
    - Re-evaluation
    - Logo items to keep you on their mind – coffee mugs, magnets, t-shirts, calendars, etc….
- You should always maintain the utmost professionalism and have materials that you can leave with the medical professionals for review or distribution.
    - Brochures
    - Business cards
    - Flyers
    - Gifts with your logo

- T-shirts
- Coffee mugs
- Lymphedema first aid kits with Band-Aids, Neosporin, and antibacterial wipes for doctors to give their patients
- Water bottles
- Towels
- Calendars

Details about making contact with the medical professionals

The List

As a Cancer Exercise Specialist you will naturally want to network with the medical professionals in your community. It will be essential that you educate them about your credentials in order to earn their trust. Some medical professionals will be reluctant to refer to you initially because they are responsible for their patients' care and they don't know you or your skill level. I suggest that you begin with your personal circle of influence:

- Primary care doctor
- Gynecologist/Obstetrician
- Dentist
- Surgeons, Doctors, Physical, Occupational, or Speech therapists that you may have used in the past

Start by creating a list of your own circle of influence and then expand it to include your clients, family, and friends' doctors, nurses, physical therapists, oncologists etc….
The professionals on this list will be more receptive to meeting or speaking with you because you are 'name-dropping.' I'm not suggesting that this is a sure thing, but it's a better place to start than cold calling.

The Letter

In the form section of this book you will find several different templates for introduction letters. Begin by sending a letter, along with your business card, to each of the people all on your list. They will be more likely to open the letter if you hand write the address on the envelope. Make a note of when each letter was mailed and follow-up with a phone call a week to ten days later. You will most likely speak to the receptionist or office manager. Introduce yourself and ask them if they received your letter. Let them know that you would like to come in before office hours or during lunch to meet with the medical professional whom you are trying to approach. It's always a good idea to let them know that you'll bring coffee and bagels, sandwiches, pizza, or whatever is suitable for that time of day.

The Meeting

You only get one chance to make a first impression – don't blow it! Someone who considers their time VERY valuable has agreed to give you ten minutes of it. Although you may be fortunate enough to have more than ten minutes, assume that that's all you have to deliver your message. You should dress professionally in a nice warm-up suit or a logo shirt with a pair of slacks. Leave the shorts and baseball hats at home. You will want to bring some time of brochure or flyer with you along with a number of business cards. If you don't have it in your budget to have a graphic designer do your brochure, you can make a perfectly acceptable brochure on your desktop publishing program. Here are a few ideas of things that you may want to include:

Benefits of Working with a Cancer Exercise Specialist

- Reduce cancer pain and fatigue
- Prevent, identify, and manage lymphedema
- Increase shoulder range of motion and correct postural deviations following mastectomy and/or reconstruction
- Increase treatment tolerance
- Prevent and/or manage Osteoporosis, Diabetes, and damage to the heart and lungs following chemotherapy, radiation, and hormonal therapies

- Return to pre-treatment levels of strength and fitness

Individualized Programming Includes:

- Postural assessment to determine muscle imbalances that may cause pain and degeneration of the joints
- Range of motion assessment to determine limitations in shoulder range of motion following cancer surgery/treatment
- Balance and core stability assessment
- Girth measurements to monitor for lymphedema
- Flexibility assessment
- Heart rate and blood pressure
- Personalized program to correct imbalances and prevent future degeneration while helping you to return to your pre-treatment level of strength and fitness

Credentials

- Your name
- Formal education or training
- Certifications held (including Cancer Exercise Specialist)
- Publications or television appearances
- Associations

Testimonials – quotes from clients and/or medical professionals

Emergency Procedures

Chest pain or pressure

- Discontinue exercise
- Have client sit down
- Advise them to see their doctor as soon as possible
- Get doctor's permission to resume exercise

Blurred vision, dizziness, lightheadedness, or faintness

- Discontinue exercise
- Have client sit or lie down on stable surface
- If the feelings don't subside, call their doctor
- Get someone to drive them to the doctor

Pain in the affected arm or shoulder (breast cancer)

- Reduce intensity of exercise
- If reducing intensity doesn't relieve symptoms, stop exercising that body part for a while
- Consult doctor or physical therapist for recommendations

Cut, scrape, burn, or bug bite in affected area of lymph node removal

- Clean area with antibacterial agent
- Cover with Neosporin and bandage

Swelling, redness, or burning of affected area

- Discontinue exercise
- Apply ice
- Elevate arm/leg and pump hand/foot
- Have patient contact doctor
- Get doctor's permission to resume exercise

Case Studies

Mary Jo:

- **33 year old female**
- **Avid exerciser**
- **Right lumpectomy (3/09)**
- **Radiation (finished 8/09)**
- **Hormonal therapy**

SAMPLE ANSWER:

Because Mary Jo is only 33 years old she is an avid exerciser, she will probably be pretty depressed because she's not only dealing with a cancer diagnosis, but it will slow her down a great deal and she will have to work back to her previous level of fitness slowly and methodically.

Following the lumpectomy on her right side, I might expect that she will have some tightness in the chest wall (depending on how much was removed). This will be compounded by the radiation to that area. Therefore, she will probably have round shoulder syndrome from the tightness across her chest. If so, I will make sure that she does exercises to stretch and open up her chest (chest fly, door stretch), avoid pressing exercises like the chest press which will shorten the muscles even more, and focus on strengthening the back and shoulder stabilizers. Because she underwent radiation over a year ago, the side-effects and fatigue should have mostly subsided by now. She will, however, be predisposed to lymphedema in her right arm because of the radiation damage to the lymph vessels and nodes. Therefore, I will take her baseline measurements of both of her arms (using the left one for comparison) and at the beginning of each workout I will take measurements of her right arm and record them for comparison. If there is swelling of ½" or more, I will refer her to her doctor and require her to come back with a medical clearance. I f the doctor prescribes a compression sleeve, I will require that she wear it for each workout. The radiation may have caused damage to her heart and lungs and I will make sure that she as the appropriate response to any aerobic activity that she does. If there is no noticeable, normal, improvement, I may have her consult with her doctor. Because she is currently on hormonal therapy, she will be thrown into instant menopause. This may add to her depression because she will not be able to have children and she will probably gain weight. The weight gain is another red flag for lymphedema because the adipose tissue retains more fluid and blocks lymphatic pathways. It will be very important to keep her weight under control through cardiovascular workouts and proper diet. I will refer her to a nutritionist who specializes in working with cancer patients, if her weight becomes an issue. She will experience all of the side effects that accompany menopause.

The program that I recommend for Mary Jo will be based on the findings from my assessment. I will focus on stretches for areas that appear tight and strengthening exercises for those muscles that are weak or opposite the tight muscles. Cardiovascular exercise will be a regular part of her activities. I will start her off slowly, even though she will want to jump right back in to her old exercise program. I will encourage her to go slowly because of the possibility of lymphedema. I will start with very light weights and only a few repetitions, and make sure there is no noticeable swelling following the workout. If she tolerates the workout, I will raise the weights and repetitions accordingly the next workout. If there appears to be some swelling, I will back off, reduce the intensity and duration and, if necessary, refer her to her doctor. I will assure her that she will be able to return to her old level of fitness in due time, but that she needs to be patient to avoid further complications.

John:

- **55 year old male**
- **Weekend warrior exerciser**
- **Total thyroidectomy w/ node dissection(8/10)**
- **Chemotherapy (currently undergoing)**

Betty:

- **65 year old female**
- **Sedentary**
- **Right modified radical w/ axillary node dissection (7/10)**
- **Currently undergoing chemotherapy**
- **30 lbs. overweight**

Larry:

- **45 year old male**
- **Marathon runner**
- **Radical retropubic prostatectomy with lymph node dissection (1/10)**
- **Hormonal therapy (ongoing)**
- **Larry wants to resume marathon training**

Charlene:

- **45 year old female**
- **Moderate exerciser**
- **Bi-lateral modified radical (8/09)**
- **Right axillary node dissection (8/09)**
- **Finished chemotherapy (2/10)**
- **Bi-lateral/contra-lateral abdominal TRAM (8/10)**
- **Hormonal therapy**

Emily:

- **34 year old female**
- **Active**
- **Right modified radical with axillary node dissection (1/09)**
- **Radiation to right side (finished 7/09)**
- **Latissimus flap w/ implant (7/10)**
- **Hormonal therapy**

Forms

Health History Questionnaire

Last name____________________________________First name________________________

Home phone__________________________Work phone_______________________________

E-mail address____

Home address____

In case of emergency contact____

Emergency contact phone____

Personal physician_______________________________phone__________________________

Date of birth____________________________________age____________

The testing and evaluation process provides information on your current level of fitness, and for the development of an individual exercise program. The fitness assessment emphasizes cardiovascular and muscular fitness as well as flexibility, range of motion, core strength, and balance. In order to get a complete and appropriate individual assessment, it is imperative that you fill out this form completely and that you don't leave out any information that could influence your individual program.

Family history – check if any of your immediate family has had:

Heart Disease ( )Mom ( )Dad ( )Grandfather ( )Grandmother ( )Brother/sister

Stroke ( )Mom ( )Dad ( )Grandfather ( )Grandmother ( )Brother/sister

Diabetes ( )Mom ( )Dad ( )Grandfather ( )Grandmother ( )Brother/sister

High blood pressure ( )Mom ( )Dad ( )Grandfather ( )Grandmother ( )Brother/sister

High cholesterol ( )Mom ( )Dad ( )Grandfather ( )Grandmother ( )Brother/sister

Other conditions/comments:____

____

If there was a documented case of heart disease, please check the age category when they first knew.

( )Under 50 years of age

( )Between 50-65 years of age

( )Over 65 years of age

Have any relatives died suddenly, without prior warning or knowledge of heart disease?

( )Yes ( )No If yes, who?__________________________Age at time of death?_____________

Personal history – check if you have had:

AIDS( ) Anemia( ) Arthritis( ) Asthma( ) Bronchitis or emphysema( )

Cancer( )____

If so, what kind?_______________________________ Surgery (type and date)___________________

____

Treatment (type and date)____

Diabetes( ) Epilepsy( ) Gout( ) Heart disease( ) Heart murmer, skipped, or rapid beats( )

High blood pressure( ) High cholesterol( ) Kidney disease( ) Lung disease( )

Phlebitis( ) Rheumatic fever( ) Stroke( ) Thyroid problems( )

Orthopedic injuries or chronic pain:

Neck( ) L shoulder( ) R shoulder( ) Cervical spine( ) Thoracic spine( ) Lumbar spine( )

L elbow( ) R elbow( ) L wrist( ) R wrist( ) L hip( ) R hip( ) L knee( ) R knee( )

L ankle( ) R ankle( ) other( )

Please explain any of the above that you have checked____________________________________

____

____

Other conditions/comments____

____

Medications

Are you currently taking any prescription medications? ( )Yes ( )No If yes, what and how much?

____

____

Are you currently taking any over-the-counter medications or vitamins? ( )Yes ( )No If yes, what and how much?____

____

Health habits

Smoking history:

Do you smoke? ( )Yes ( )Quit ( )Never

What do/did you smoke? ( )Cigarettes ( )Cigars ( )Pipe

How much did/do you smoke a day?____

How long have you been smoking?_____________If quit, when?____________________

Exercise habits

Do you engage in physical activity? ( )Yes ( )No

What kind?____

How hard? ( )Light ( )Moderate ( )Hard How often?__________________________

Did your past exercise habits differ from what you are doing now? ( )Yes ( )No

What kind of exercise did you do in the past?___________________________________

____

How hard? ( )Light ( )Moderate ( )Hard How often?__________________________

Is your occupation ( )Sedentary ( )Active ( )Heavy work
Explain:____

Do you experience discomfort, shortness of breath, or pain with exercise? ( )Yes ( )No

If yes, what type of exercise/symptoms?______________________________________

____

Nutritional behavior

Do you consider yourself overweight? ( )Yes ( )No

If yes, how long have you been overweight?___________________________________

How many meals do you typically eat per day?__________________________________

How often do you eat outside the home?____________________________per week

How much of the following do you consume?

____________cups of caffeinated coffee or tea per day

____________glasses of caffeinated soda per day

____________glasses of beer per day (12oz. = 1 unit)

____________glasses of wine per day (4 oz. = 1 unit)

____________glasses of liquor per day (11/2 oz. = 1 unit)

____________units of alcohol per week (see above for unit equivalent)

Stress

Do you consider your day stressful? ( )Yes ( )No

What is the nature of your stress?____

How many hours do you sleep a night?____________Is your sleep sound? ( )Yes ( )No

Do you practice any form of meditation? ( )Yes ( )No If so, what?______________________________

____

What is your preferred training schedule? (days/hours of availability)

____

Personal Assessment

Name____ Date____________________

RHR____________________ RBP____________________ Weight__________________

Postural Assessment:

	Yes	No	Severe	Moderate	Minimal	Internally	Externally	L	R
Forward Head									
Tilted Head									
Kyphosis									
Rounded Shoulders									
Elevated Shoulders									
Winged Scapula									
Scoliosis									
Lordosis									
Hips Level									
Hips Tilted									
Hips Rotated									
Knees Rotated									
Feet Rotated									
Tight Hip Flexors									
Tight Quads									
Tight ITB									
Tight Calves									
Tight Peroneals									
Feet Supinate									
Feet Pronate									

ROM Assessment:

	Active Left	Active Right	Passive Left	Passive Right
Shoulder Flexion				
Shoulder Extension				
Shoulder Abduction				
Shoulder IR				
Shoulder ER				
Hip Flexion				
Hip Extension				
Hip Abduction				
Hip Adduction				
Knee Flexion				
Knee Extension				
Hamstring				

Girth Measurements:

	Right	Left
Neck		
Chest		
Waist		
Hips		
Upper Thigh		
Mid-Thigh		
Knee		
Calve		
Ankle		
Wrist		
Mid-Ulna		
Elbow		
Mid-Humerus		

Core and Balance Assessment:

	Left	Right
Stork		
All Fours		
Plank		

Crunches ___________________ **Push Ups** ___________________

Body Fat Women: __________Thigh __________Tricep __________Suprailium

__________**Total %**

Body Fat Men: __________Thigh __________Abdomen __________Chest

__________**Total %**

Quality of Life Questionnaire

1) How would you rate your overall satisfaction with life?
○ poor ○ below average ○ good ○ very good ○ excellent

2) How would you rate your current health and well being?
○ poor ○ below average ○ good ○ very good ○ excellent

3) How often do you get sick or go to the doctor?
○ once a week ○ twice a month ○ once a month ○ once every six months ○ once a year

4) How would you rate your past fitness level?
○ poor ○ below average ○ good ○ very good ○ excellent

5) How would you rate your current fitness level?
○ poor ○ below average ○ good ○ very good ○ excellent

6) How would you rate your perceived body image?
○ poor ○ below average ○ good ○ very good ○ excellent

7) How would you rate your current energy level?
○ poor ○ below average ○ good ○ very good ○ excellent

8) How would you rate your current ability to enjoy activities?
○ poor ○ below average ○ good ○ very good ○ excellent

9) How would you rate your current mobility?
○ poor ○ below average ○ good ○ very good ○ excellent

10) How would you rate your current level of pain?
○ no pain ○ manageable pain ○ chronic pain ○ unbearable pain

11) How would you rate your past eating habits?
○ poor ○ below average ○ good ○ very good ○ excellent

12) How would you rate your current eating habits?
○ poor ○ below average ○ good ○ very good ○ excellent

13) How would you rate your current ability to perform activities of daily living (bathing, grooming, dressing, cooking, cleaning…)?
○ poor ○ below average ○ good ○ very good ○ excellent

14) How would your rate your current ability to perform work-related tasks?
○ poor ○ below average ○ good ○ very good ○ excellent

15) How would you rate your sleep at night?
○ poor ○ below average ○ good ○ very good ○ excellent

16) How would you rate your current mood?
○ depressed ○ mildly depressed ○ content ○ happy ○ very happy

Contract for Personal Fitness Services From The Cancer Exercise Training Institute

Our goal at the Cancer Exercise Training Institute is to provide a client personalized fitness training including cardiovascular conditioning, weight training, flexibility, and range of motion, as well as postural deficiencies and corrections. The Cancer Exercise Training Institute is dedicated to the overall success of each client's personal health and fitness program.

1. **Payment** – payment for personal training services will be billed up front prior to the date of the first training session.

2. **Sessions** – each session will be one hour in length. Client agrees to be on time for each session. The session starts and ends when scheduled. If the client is late, the session will end as scheduled despite his/her lateness.

   Once a session is scheduled between the client and Cancer Exercise Training Institute, it is the client's responsibility to either attend the scheduled session or provide Cancer Exercise Training Institute with at least twenty-four hours notice of cancellation. If the client cancels within twenty-four hours of the session, the client shall be responsible for the session.

3. **Physician's Approval** – Client must fill out and submit the health history questionnaire form and medical clearance form along with this liability release, and made part of the agreement together with any documents, reports, or other information provided by client's physician or doctor.

   Cancer Exercise Training Institute has the right to refuse any session to a client at any time based upon his/her physical condition.

4. **Waiver** – the client is aware that Cancer Exercise Training Institute sessions include strength, flexibility, and aerobic exercise, which may be potentially hazardous activities. The client also acknowledges that these activities involve a risk of injury and death. Accordingly, the client voluntarily consents to participate in Cancer Exercise Training Institute sessions, and assumes the acknowledged risks involved.

   The client hereby waives, release, and forever discharges Cancer Exercise Training Institute, its officers, agents, employees, representatives, and executors, from any and all responsibilities or liability from injuries or damages resulting from participation in a class. Client also agrees that Cancer Exercise Training Institute, its officers, agents, employees, representatives, and executors shall not be liable for any claim, demand, cause, or action of any kind whatsoever for, or on the account of death, personal injury, property loss, or damage resulting from participation in any session.

________________________________ ____________

Client Signature Date

MEDICAL CLEARANCE FORM

Dear Doctor:

_____________________________has applied for enrollment in the fitness testing and exercise programs at the Cancer Exercise Training Institute. The fitness testing program involves a submaximal test for cardiorespiratory fitness, sit and reach flexibility test, arm girth measurements, postural assessment, shoulder ROM test, muscular strength and muscular endurance tests. The exercise program is designed to start with basic stretching and flexibility exercises along with the use of light resistance to increase upper and lower body strength. The client is evaluated every six weeks to reassess their status and determine whether or not to advance them to the next level of difficulty. The program will take the client through various levels of increasing difficulty. All of our trainers are certified by the Cancer Exercise Training Institute as a Cancer Exercise Specialist. Thus they have undergone thorough and intensive training in working with the special needs of cancer survivors.
By completing the form below, however, you are not assuming any responsibility for our administration of the fitness testing and/or exercise programs. If you know of any medical or other reasons why participation in the fitness testing and/or exercise programs by the applicant would be unwise, please indicate so on this form. If you have any questions about the program, please don't hesitate to call us at:

REPORT OF PHYSICIAN

____

__________ I know of no reason why the applicant may not participate

__________ I believe the applicant can participate, but I urge caution because: _______________

____

__________ The applicant should not engage in the following activities: __________________

____

__________ I recommend that the applicant not participate.

Physician signature ________________________________ Date ________________________

Address ____ Phone ______________________

City and State ______________________________________ Zip ________________________

CES Background Information

The Cancer Exercise Specialist Advanced Qualification was developed for health and fitness professionals seeking to attain a higher level of mastery and work with cancer patients in post-operative exercise and quality of life programs. Following the workshop, there will be a take-home examination that candidates must pass in order to receive their advanced qualification.

The course consists of the following:

- Exercise implications and contraindications for 25 types of cancer, surgery, reconstruction, and treatments
- Breast reconstruction and contraindications to exercise
- Preventing and identifying lymphedema
- Cancer related pain
- Mental and physical fatigue during cancer treatment
- Conducting a postural and range of motion assessment, interpreting the results, and creating an individualized exercise program
- Working with the medical professionals – Marketing strategies
- Adapting your equipment to various cancer procedures to avoid injury and maximize benefits
- Conquering Cancer With Nutrition – Dr. Glenn B. Gero

This program is based, in part, on "Essential Exercises for Breast Cancer Survivors" and the EM-POWER Program, originally developed by CES professional and author Andrea Leonard with the following medical advisory board:

Dr. Theodore Tsangaris, Jr. - Former Chief of Breast Surgery at the Georgetown University Medical Center, Chief of Breast Surgery at Johns Hopkins Medical Center
Dr. Katherine Alley - Chief of Breast Surgery at Suburban Hospital
Dr. Shawna Willey - Former Chief of Breast Surgery at the George Washington University Medical Center, Chief of Breast Surgery at the Georgetown University Medical Center
Dr. Richard Flax - Breast Surgeon at the Columbia Hospital for Women
Jean Lynn, R.N. - Oncology Nurse, Mammocare Director at the George Washington University Medical Center
Rosalie Begun, P.T. - Physical Therapist at Begun Physical Therapy
Amy Halverstadt, M.S. - Exercise Physiologist, Co-author of "Essential Exercises for Breast Cancer Survivors."

Andrea Leonard has been conducting the CES workshops nationwide since 1995. She is Co-author of "Essential Exercises for Breast Cancer Survivors," Founder and President of The Cancer Exercise Training Institute, and a continuing education provider for The American Council on Exercise and The National Academy of Sports Medicin. Andrea is certified as a Special Populations expert by The Cooper Institute, as a Personal Trainer by the American Council on Exercise and The American College of Sports Medicine, as an Optimum Performance Trainer by The National Academy of Sports Medicine, and as a Strength and Conditioning Coach by The National Sports Professionals Association.

SAMPLE PRESS RELEASE

Contact:
Cancer Exercise Training Institute
3269 Forest Ct.
West Linn, OR 97068
Contact: Andrea Leonard, (503) 502-6776

MEDICAL AND FITNESS PROFESSIONALS NATIONWIDE HELP CANCER PATIENTS TO PREVENT AND REVERSE THE RAVAGES OF CANCER SURGERY AND TREATMENT

Andrea Leonard, CANCER SURVIVOR, Co-Author of "Essential Exercise for Breast Cancer Survivors," Founder of the Breast Cancer Survivors Foundation and President of The Cancer Exercise Training Institute, has been conducting two-day Cancer and Exercise Workshops for health and fitness professionals nationwide since 2001. Students in her classes are naturopathic doctors, nurses, physical therapists, personal fitness trainers, and Yoga and Pilates instructors. Thanks to Andreas' comprehensive Cancer Exercise Specialist Advanced Qualification, there are now thousands of professionals covering the United States, British Columbia, The Netherlands, Greece, Australia, Singapore, Israel, and Puerto Rico. The Cancer Exercise Specialist is truly a pioneer in the practice of cancer recovery and rehabilitation from the debilitating side-effects of cancer surgery and treatment. For more information on upcoming workshops you can visit the website at www.thecancerspecialist.com.

"This Advanced Qualification is to cancer, what Cardiac Rehabilitation has been for heart attack victims," declares Andrea Leonard, Cancer Exercise Training Institute President, and herself a twenty-five year cancer survivor.

Following cancer surgery and treatment there are numerous physically debilitating side-effects that cancer patients have become accustomed to living with. We want to make it known that cancer survivors no longer have to accept the fact that they have limited range of motion, poor posture, neck and back pain, lymphedema, chronic fatigue etc… following their treatment. We have the answers and can help to reverse many, if not all, of the agonizing problems that chronically plague cancer survivors nationwide. In addition to the many public and private facilities that now have Cancer Exercise Specialists; we have certified professionals at Hospitals and hospital-based wellness facilities across the country.

"After 24 years in the fitness industry, I once again feel as I am making a difference in my clients' lives. It's so empowering to work with people that do NOT take everyday for granted and are motivated to improve every aspect of their remaining time. As selfish as this may sound, I feel that I am getting as much if not more out of this than my clients."

Robert Reed III, Certified Personal Trainer and Cancer Exercise Specialist, San Antonio, TX

Contact Andrea Leonard at (503)502-6776 for more information.

Sample Letter to Doctor When Working with Their Patient

September 1, 2006

Dear Dr. Rubin,
My name is Andrea Leonard. I live in West Linn and am a Cancer Exercise Specialist. I am Co-author of "Essential Exercises for Breast Cancer Survivors," and president of The Cancer Exercise Training Institute. Today I had the good fortune of meeting with one of your breast cancer patients, Deborah Searfus. I conducted a comprehensive evaluation of Deborah's ROM, arm girth measurements, and a postural assessment to determine muscle imbalances that exist prior to her mastectomy; so that we have a base for comparison later. Following her mastectomy, we will repeat this process. I have enclosed a pre-operative report for you, listing all of my findings along with a very brief synopsis of my intended exercise programming. She already has limited abduction in her right arm, possibly having something to do with the tumor location and depth. Flexion is also slightly limited. Her other measurements are within normal range. If you have any questions or concerns, or if I can be of assistance to any of your other patients, please don't hesitate to contact me at (503) 502-6776.

If you would ever like to have coffee, I would love to be able to share more of what I do with you.

Respectfully,

Andrea Leonard

Sample Marketing Letter to Doctor

September 1, 2006

Dear Dr. Nakamura,
My name is Andrea Leonard and I am a Cancer Exercise Specialist residing in West Linn, OR. I wanted to take this opportunity to introduce myself and offer my services to your patients. I work individually with clients, helping to minimize the side effects of treatment, reverse postural and range of motion issues that arise from surgery and reconstruction, and focus on awareness and prevention of lymphedema. I have enclosed a recent article from the Oregonian which explains what I do and what my background is. The following is a summary of my qualifications:

- Adjunct member of the American Council on Exercise and National Academy of Sports Medicine Faculty
- National presenter of accredited Cancer and Exercise Workshop
- Co-author of "Essential Exercises for Breast Cancer Survivors"
- Author of "The Cancer and Exercise Handbook"
- President of Leading Edge Fitness
- President/Founder of The Cancer Exercise Training Institute
- B.A. University of MD, 2000
- Certified Personal Trainer, Conditioning Specialist, and Special Populations Expert:
    - American College of Sports Medicine
    - National Academy of Sports Medicine
    - American Council on Exercise
    - National Sports Professionals Association
    - Cooper Institute

I would welcome the opportunity to have coffee with you and answer any questions that you may have. I can also provide you at that time with references; if you would like to see them. I hope to have the opportunity to serve your patients in the near future.

In Health,

Andrea Leonard

Biography

Andrea Leonard earned her BA from the University of Maryland in 1990 and continued her education to pursue a career in fitness. She is certified as a Conditioning Specialist by the National Sports Professionals Association (NSPA), Personal Trainer by the American Council on Exercise (ACE), Health and Fitness Instructor by the American College of Sports Medicine (ACSM), Optimum Performance Trainer by the National Academy of Sports Medicine (NASM), and as a Special Populations Expert by The Cooper Institute. Andrea is also an adjunct faculty member for ACE & NASM.

A cancer survivor herself at age eighteen, Andrea watched her mother struggle through two breast cancer diagnoses over the course of twenty years. Watching the pain and suffering that her mother needlessly endured, "frozen shoulder," narcotics addiction - due to chronic pain from an axillary node dissection, negative postural deviations from bi-lateral mastectomies and attempted reconstruction, staph infections, osteoporosis, etc....Andrea was bound and determined to help the millions of cancer survivors out there looking for answers on how to take back control of their life.

Andrea, along with a team of medical professionals from Georgetown University Hospital, Johns Hopkins Medical Center, and The George Washington Medical Center, spent three years working on her first book, "Essential Exercises for Breast Cancer Survivors. She then moved from her hometown of Washington, D.C. to West Linn, Oregon. Andrea re-opened the doors of Leading Edge Fitness Consultants, a fitness consulting and Education Corporation - in business since 1993, in West Linn, Oregon. Determined to do more, Andrea founded the Breast Cancer Survivors' Foundation, a 501(c)3 non-profit organization to help breast cancer survivors and their families.

Wanting to spread her knowledge about exercise, and the important role it plays in cancer recovery, Andrea wrote the "Cancer and Exercise Handbook," developed a curriculum for health and fitness professionals, and began teaching national workshops to spread the gospel to others. As a continuing education provider for The American Council on Exercise and The National Academy of Sports Medicine, Andrea teaches a two-day Advanced Qualification for medical doctors, naturopathic doctors, nurses, physical and occupational therapists, massage therapists, personal trainers, group exercise instructors, and yoga and Pilates instructors wanting to expand their scope of knowledge. Andrea is currently the president of The Cancer Exercise Training Institute.

Andrea has been a guest speaker for several international and national fitness organizations, including TSI - Town Sports International Summit, Northwest Strength & Conditioning Clinic, Medical Fitness Association Annual Conference, and IHRSA – International Health, Racquet, & Sportsclub Association Annual Conference, as well as appearing regularly on AM Northwest. Andrea has been interviewed by The Oregonian, The Portland Tribune, The Lake Oswego Review, The West Linn Tidings, The Washington Post, The LA Times, The Dallas Morning News, The New York Times and more.

When she is not on the road Andrea lives in West Linn, Oregon with her two children; Dylan and Marin. You can learn more information about The Cancer Exercise Training Institute at www.thecancerspecialist.com

Bibliography

1) Halverstadt and Leonard, Essential Exercises for Breast Cancer Survivors, the Harvard Common Press, Boston, MA, 2000.

2) Mallon, Brenda, Creative Visualization with Colour, Element Books, Inc., Boston, MA, 1999

3) Love, Susan, Susan Love's Breast Book, Addison-Wesley Publishing Company, Menlo Park, CA, 1995

4) LaTour, Kathy, The Breast Cancer Companion, Avon Books, NY, NY, 1993

5) Komarnicky and Rosenberg, What To Do If You Get Breast Cancer, Little, Brown, and Company, Boston, MA, 1995

6) Weiss and Weiss, Living Beyond Breast Cancer, Times Books, NY, NY, 1997

7) Austin and Hitchcock, Breast Cancer - What You Should Know (But May Not Be Told) About Prevention, Diagnosis, and Treatment, Prima Publishing, Rocklin, CA
1994

8) Teeley, Peter and Bashe, Phillip, the Complete Cancer Survival Guide, Broadway Books, NY, NY, 1998

9) Burt, Jeannie and White, Gwen, Lymphedema – A Breast Cancer Patient's Guide To Prevention and Healing, Hunter House, Berkeley, CA, 1999

10) Buckman, Dr. Robert, What You Really Need to Know about Cancer,
Johns Hopkins University Press, Baltimore & London, 1995, 1997

11) Morra, Marion and Potts, Eve, Choices, Harper Collins, NY, NY, 1980, 1987, 1994, 2001

12) Schneider, Dennehy, and Carter, Exercise and Cancer Recovery, Human Kinetics, Champaign, IL, 2003

13) Clark, Michael A., NASM Certified Personal Trainer Optimum Performance Training Manual-2nd Edition, U.S.A. 2004

14) Cipriano, Joseph J., Photographic Manual of Regional Orthopaedic and Neurological Tests- 4th Edition, Lippencott, Williams, and Wilkins, 2002.

15) Norkin, Cynthia C., and White, D. Joyce, Measurement of Joint Motion – 4th Edition, E.A. Davis Company, Philadelphia, PA, 2009

16) Page, Phil, Frank, Clare C., and Lardner, Robert, Assessment and Treatment of Muscle Imbalance – The Janda Approach, Human Kinetics, Chmapaign, IL, 2010